T0278168

# Pure Human

# ALSO BY GREGG BRADEN

**Books**

*The Science of Self-Empowerment*

*Resilience from the Heart*

*The Turning Point*

*Deep Truth*

*The Divine Matrix*

*Fractal Time*

*The God Code*

*The Isaiah Effect**

*Secrets of the Lost Mode of Prayer*

*The Spontaneous Healing of Belief*

**Audio Programs**

*An Ancient Magical Prayer* (with Deepak Chopra)

*The Turning Point*

*Awakening the Power of a Modern God*

*Deep Truth*

*The Divine Matrix*

*The Divine Name* (with Jonathan Goldman)

*Fractal Time*

*The Gregg Braden Audio Collection**

*Speaking the Lost Language of God*

*The Spontaneous Healing of Belief*

*Unleashing the Power of the God Code*

**DVDs**

*The Science of Miracles*

All available from Hay House except items marked with an asterisk.

Please visit:

Hay House USA: www.hayhouse.com®
Hay House Australia: www.hayhouse.com.au
Hay House UK: www.hayhouse.co.uk
Hay House India: www.hayhouse.co.in

# Pure Human

## THE HIDDEN TRUTH OF OUR DIVINITY, POWER, AND DESTINY

GREGG BRADEN

HAY HOUSE LLC
Carlsbad, California • New York City
London • Sydney • New Delhi

***Published in the United States by:*** Hay House LLC: www.hayhouse.com®
***Published in Australia by:*** Hay House Australia Publishing Pty Ltd: www.hayhouse.com.au
***Published in the United Kingdom by:*** Hay House UK Ltd: www.hayhouse.co.uk
***Published in India by:*** Hay House Publishers (India) Pvt Ltd: www.hayhouse.co.in

*Cover design:* Jordan Wannemacher

*Interior design:* Bryn Starr Best

*Interior photos/illustrations:* Images on pages 45 and 46 courtesy of Shutterstock. Image on page 102 adapted from Shekh M. Mahmudul Islam, "Performances of Multi-Frequency Voltage to Current Converters for Bioimpedance Spectroscopy," *Bangladesh Journal of Medical Science* 5, no. 1 (October 2012).

*Indexer:* J S Editorial, LLC



Cataloging-in-Publication Data is on file at the Library of Congress

Hardcover ISBN: 978-1-4019-4936-5
E-book ISBN: 978-1-4019-4937-2
Audiobook ISBN: 978-1-4019-6284-5

10 9 8 7 6 5 4 3 2 1
1st edition, January 2025

Printed in the United States of America

This product uses responsibly sourced papers and/or recycled materials. For more information, see www.hayhouse.com.

All the powers in the universe are already ours. It is we who have put our hands before our eyes and cry that it is dark.

— Swami Vivekananda (1863–1902),
Indian Hindu monk and philosopher

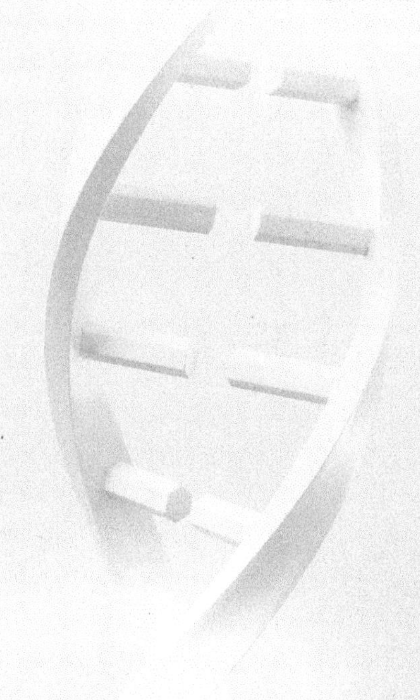

# Contents

# Preface

We humans are an ancient and mysterious form of life. We're the unlikely convergence of invisible thoughts, emotions, and imagination woven into the fabric of tissue, bone, and blood that make possible our choices, and the consequences of our choices, each and every day of our lives.

In our enigmatic state of being human, we search for others of our kind to share our joy, our love, and our dreams, as well as to ease the hurt, the fear, and the suffering arising from our sensual journey on Earth.

This book is written to provide a lens through which to view our time in history, as well as to remind us that the awakening of our divinity is the reason for our journey.

# Introduction

There are moments in human history when we make choices that can irreversibly change the world, and our lives, in ways that we may regret in the future. Today is one of those moments.

We now have at our fingertips the technology to alter ourselves—to rewrite the code of our DNA and the neural networks that define us—in ways that, once implemented, can never be reversed, and will forever change what it means to be human. We also live in a time when we're being told that we need to make precisely such changes to become the best version of ourselves and be successful in the world.

It's the combination of these life-altering technologies merged with the narrative of our powerlessness that makes our time in history so vastly different from times past.

The technologies humans have developed since we first appeared on the planet 200,000 years ago have always been limited to influencing our relationship with the world *around* us. For example, hundreds of years ago we made the choice to burn wood and coal to power our civilization, and more recently petroleum, as the primary fuel that would heat, feed, and power the nations of the world. We chose to apply the knowledge of how to split the atom—a technology that is relatively benign in itself—to build the most destructive weapons in human history. Furthermore, we've chosen to use the threat of those weapons to shape the politics and economies of the world.

We've chosen to solve our differences in religion, politics, and government by using advanced intelligence gathering, privacy-shattering surveillance, and high-tech warfare to achieve desired outcomes both internationally and domestically.

As powerful and destructive as these choices have been, however, their impact has been limited to temporary, and mostly reversible, changes in the world around us. That is, until now. And this is precisely why the crossroads we've reached is dangerously unsettling.

*As you're reading this book, policies are already being written and the technology is already being implemented that is permanently changing the world within us.* The technologies that we're accepting as individuals and families today, as well as the legal framework governing their use, are already affecting how our brains function. They're already affecting how well our immune systems can respond to newly emerging viruses, bacteria, and other contagions in the world. They're already changing the emotional bonds between friends and between parents and their children. They're already altering our ability to detect the subtle energy that we use subconsciously to communicate with one another as well as other forms of life.

Perhaps most concerning of all, they're already changing the way we feel, share emotions, and preserve the values that we cherish as individuals, families, and communities—the very essence of society.

The irony of using the new technology on a universal basis to modify our bodies is that the capabilities of our natural biology already equal, and in some cases even exceed, the capabilities of the artificial technology we're told we need to enhance our human condition.

The next steps that we take regarding artificial intelligence (AI) and the merging of our natural bodies with technology by implanting computer chips in the brain and injecting gene therapies into our blood will set the course that determines how much of ourselves we preserve and how much of our humanness we forever give away to technology.

The choice is ours and we're already making it.

**PURE HUMAN TRUTH 1:** For the first time in human history, we're implementing technology that irreversibly changes our bodies on a biological level.

If we continue on the current technological path, guided by the current trends in thinking, by the year 2030 we will have made the ultimate choice: We'll be well on our way to one of two kinds of societies. We will either be locked into a "futuristic" society of human-machine hybrids where we've traded our cherished qualities of intuition, empathy, and creativity and the soul-stirring bonds of love, intimacy, and sexual conception for the convenience of AI that creates our music, poetry, and art, and virtual realities that replace genuine relationships and human contact.

Or we will awaken to the deep truth of our extraordinary yet largely untapped natural human potentials and, for the first time as a species, discover what it means to be fully human.

**PURE HUMAN TRUTH 2:** By the year 2030, we will either have awakened to the truth of our untapped human potential, or we will be locked into a society of hybrid humans that has engineered away our powers of creativity, emotion, empathy, and intuition.

We can only make the informed and healthy choices that determine which of these paths we want if we know who we are as humans and understand the full magnitude of what's at stake.

## WHAT KIND OF WORLD DO WE WANT?

American inventor and futurist Ray Kurzweil has a long history of successfully predicting where the trends of technology will lead, and what kind of world they will lead to. In 1990, for example, he accurately predicted the widespread adoption of computers and the use of the Internet in average households, as well as the emergence of driverless electric cars controlled by AI.

As early as 2005, Kurzweil recognized that the human-machine merger we know today as *transhumanism* was on the horizon. Based upon the trajectory of social norms and technology of that time, it was the logical convergence of the quest for human immortality along with advances in robotics and super-small computers. He also recognized that, unless something changed in the way we thought about ourselves, the transition from humans to transhumanistic human-machine hybrids would be completed even sooner than many people expected. It would happen, he said, and it would be embraced in plain sight as "progress" and the next step in our evolution. Furthermore, it would happen without the long-term implications being made clear to the general public.

During a 2013 interview, Kurzweil reiterated just how quickly the movement toward human-machine interfaces was progressing, stating, "When you talk to a human in 2035, you'll be talking to someone that's a combination of biological and non-biological intelligence."[1]

If Kurzweil was right about this, ours may well be the last generation of pure humans on Earth. We have only a few years left to determine the future course of our relationship to advanced robotics and AI.

What sets today apart from similarly critical moments in our past is that if we make the wrong choices now, what we stand to lose is monumental, unprecedented, and irreversible. *It's us.* We stand to lose ourselves, our humanness—and the unique qualities that set us apart from other forms of life.

Ultimately, we stand to lose the part of us that has been the root of history's greatest triumphs, the vessel holding our deepest secrets and the most prized aspects of our human nature. We stand to lose the source of our imagination, intuition, innovation, and creativity. What's at stake is our divinity—the part of us that allows us to rise above our circumstances and become more than any limitations and expectations that we've accepted for ourselves in the past.

**PURE HUMAN TRUTH 3:** We stand at the precipice of giving away our humanness—the biological bridge to our divinity.

Ironically, the potential to lose our divinity is not an accident. It's not the consequence of "progress" gone awry. Rather, it's the goal of a movement that began in the 20th century to transform humans into a new form of life that is beyond human—a transhuman or post-human species. The stated goal of this transhumanism is to achieve immortality and incorporate the logic, speed, and efficiency of computers and AI into the human body. Advocates of this movement believe that transhumanism is the next step in human evolution.

The steep price that we will pay for such achievements is the creation of emotionless individuals, driven by efficiency, logic, and algorithms, who no longer experience what transhumanists view as the human "flaws" of uncomfortable emotions, such as grief, suffering, broken love, and loss.

In a transhumanistic world, the inspiration that has compelled some of the greatest artistic achievements of our species, such as the romantic tragedy of Shakespeare's *Romeo and Juliet*, the profound poetry of Rumi, the intensity of Beethoven's "Moonlight Sonata," the passion of Led Zeppelin's "Stairway to Heaven," and the natural beauty of Leonardo da Vinci's *Mona Lisa*, will become an artifact of our past. A distant memory of another time for our descendants.

Ultimately, transhumanists strive for the complete merging of humans and machines into a unified digital landscape (literally a digital matrix) called the *singularity*. In the singularity, we humans are integrated into the Internet of everything, digitally entangled in a vast complex of information that's managed and controlled through AI-driven algorithms. Nearly undetectable nanotransmitters, some of which are already used for medical applications today, will circulate in our bloodstreams, where they allow our vital signs and intimate bodily functions to be monitored and documented on remote servers.

The same technology will serve a dual purpose, however, as it will also be monitoring our everyday choices and habits, including what we eat, what we purchase, how and where we travel, and even how we feel about one another, while assigning consequences when we step outside of the socially accepted norms.

Kurzweil believes this human-tech merger is occurring today, and he has identified a date for when we can expect it to be the predominant form of life in our civilization: "I set the date for the singularity—representing a profound and disruptive transformation in human capability—as 2045."[2]

## THE MISSING LINK

The path that Kurzweil and others foresee sounds like something from a really bad, dystopian science fiction movie, with a theme that we've all seen in such movies before. For most of us, when we see the dark-themed movies or read about the gloomy predictions, we feel an instant resistance, and our guts scream, "No!" This isn't a future we would want for ourselves or for our children.

The reason for our resistance is because, deep within ourselves, we sense that it's all wrong. It's wrong to allow high-tech gadgets to intrude upon the sanctity of our natural bodies. It's wrong to replace our natural ability for thought, imagination, and creativity with the streamlined and emotionless efficiency of AI and computer chips. The entire premise is wrong because we know instinctively that there is something very special about our existence. Although it is seldom acknowledged or spoken about in our culture, we have a force within us that we stand to lose access to if we give ourselves away to machines.

And although we may not be able to quite place our finger on what the "special something" is that we feel, we know that we embody an ancient gift, and a sacred trust, that is ours to preserve and protect. As dysfunctional as the world and the people in it may appear to us today, we still sense that there is something special about our existence that shines through the dysfunction. It is this something that's worth preserving. It's this rare gift within us that this book is about.

From our most ancient and cherished spiritual traditions to the best science of the modern world, new discoveries continue, year after year, to reveal ever more ways that we humans are unique, and even more special than we've ever allowed ourselves to believe. It's through the nearly godlike powers of our emotions, our empathy, our intuition, our forgiveness, and our innovation that we already embody the very powers of creation that the transhumanists believe only human-machine hybrids can achieve.

As we'll discover in the chapters that follow, human cells and neurons actually out-power, out-perform, and out-adapt the limited scalability of hardwired computer chips and the limited algorithms of AI.

**PURE HUMAN TRUTH 4:** In many respects, human cells and specialized neurons are superior in performance, scalability, and adaptability to hardwired computer chips and the limited algorithms of AI.

While the technology that transhumanists are actively striving to integrate into our lives may be seductive, these innovations actually mimic the natural functions that the living tissues of our bodies already perform.

## GOOD AND EVIL

To fully embrace the magnitude of what's at stake when it comes to risking our humanness while merging our bodies with machines, we need to acknowledge the conflict that has raged since the beginning of our existence on Earth—the timeless battle between good and evil.

Although the battle itself is ancient, it's also current. It's playing out on the global stage, and in our lives today. And while it's not always easy to talk about, and those who do talk about it are often chastised, shamed, and "canceled" by members of their social circles for doing so, the battle between good and evil is alive, present, and very real. It's at the core of the conflict, war, disease, and social breakdown that we see in families, communities, and nations today. *Evil is also at the core of the transhumanist movement that wants to replace the natural biology of our bodies with machines.*

As we'll see in Chapter 1, the divinity that frees us from the vulnerability of fear is only possible through the power of the natural human body. From this perspective, the movement to replace our natural bodies with synthetic intelligence and edited genes is ultimately a movement that separates us from the divinity that empowers us to express the depths of our humanness.

We stand at the threshold of giving away our 2,000-century heritage of human potential and the remarkable abilities that we are only beginning to recognize. We owe it to ourselves to awaken those potentials in our lives and to understand who we are in their presence before our destiny is forever lost to "progress" someone else's idea of high-tech evolution.

**PURE HUMAN TRUTH 5:** We owe it to ourselves to recognize the deep truth of what it means to be human before we give ourselves away to the technology now being proposed by the transhumanist movement.

The purpose of this book is to remind us all of just how powerful, precious, and sacred we are. I've written *Pure Human* to offer us a sense of direction and peace, as together we chart the course of our future and our destiny. Our humanness is worth preserving. If you've ever wanted to think differently about yourself and your untapped potentials, but were reluctant to do so due to a lack of evidence revealing those potentials, the pages that follow are for you.

It is my hope that this book creates for you a renewed sense of appreciation for the honor and privilege, each day of your life, of carrying the blueprint of our humanness in every cell of your body.

Written with love,

GREGG BRADEN
SANTA FE, NEW MEXICO

CHAPTER ONE

# We Are the Prize

## The Battle for Our Humanness

We are a blend of dust and divinity.

— Huston Smith (1919–2016),
American scholar of religious studies

An ancient battle is playing out before our very eyes.

You won't see this battle reported on the six o'clock news or read about it the morning newspaper. It won't be discussed during a televised presidential debate or White House press briefing. The battle itself is disguised in the affairs that drive the events of our world. It's constant and relentless.

While it's easy to define this battle broadly as the timeless struggle between good and evil, the battle is ultimately for something that lies beyond these familiar polarities. The objective of this battle is the domination of a powerful force that lives within us. Within each of us. This force is the key to experiencing joy, success, and healing in our everyday lives as individuals. It's the force that frees us from the fear that keeps us feeling small, insignificant, and powerless. Ultimately, it is this force that will decide the future and destiny of our entire species.

The force that I'm talking about can be encapsulated in a simple word: *divinity*. More precisely, it is the power of our *humanness* that allows us to express our divinity in everyday life that is the object of the ancient battle.

## WHAT IS DIVINITY?

In ancient Hindu traditions, there is a single word for a mysterious force that has no direct equivalent in the English language. That word is *atman*, and it means the "eternal divine." Although the word *atman* is not typically used in Western traditions, the divinity that it describes is, and will be familiar to most Westerners.

And while the meaning of the word *divinity* itself has historically been associated with religion, religious education, or some kind of spiritual practice, a deeper look into its definition reveals something that may be unexpected, and to some people surprising. A closer examination of the word reveals an empowering relationship between us and our current world of extremes.

A contemporary definition of *divinity* states:

> *A divine force or power. Powers or forces that are universal, or transcend human capacities.*[1]

When we examine this definition, two facets of good news are revealed. First, *transcending* a situation is more than simply surviving the situation. It is rising above the situation to triumph over our challenges. In doing so, we become something more—something greater—than the versions of ourselves that faced the challenges in the past.

Second, *human capacities* describe what we accept as our abilities, and the limits of those abilities, both today and in the past. It's often the case, however, that the limits we are accepting for ourselves are not true limits at all. Rather, they are *perceived* limitations that we've been conditioned and indoctrinated to accept for ourselves.

> **PURE HUMAN TRUTH 6:** Divinity is defined as powers, or forces, that transcend perceived limitations.

Through our family beliefs, social practices, and the information presented to us in our classrooms and textbooks, we are each indoctrinated to accept barriers to our performance, our ability to heal, and the power of our imagination that often aren't real. A perfect example of this is our ability to achieve higher brain states, leading to greater levels of consciousness and perception, and to do so intentionally and on demand.

## WHEN LIMITS ARE NOT LIMITS

As late as the 1990s, university textbooks and scientific papers acknowledged only four states of awareness that the human brain can achieve. These levels of awareness—encompassing various kinds of awake and asleep consciousness—were based upon the readings of brain waves from electroencephalographs (EEGs) that measure neural activity electrically. Briefly, from the lowest-frequency brain state documented to the highest that was recognized at the time, these states and their associated frequencies were: the delta state of deep sleep (1.5 to 4.0 Hz), the theta state of REM sleep and deep meditation (5.0 to 8.0 Hz), the alpha state of wakeful relaxation (9.0 to 14.0 Hz), and the beta state of active alertness (15.0 to 40.0 Hz).[2] Within these four acknowledged states, the maximum frequency for the human brain was believed to be the 40 Hz of the beta state.

Scientists at the time were absolutely certain that the human brain could not sustain frequencies beyond this range, and that 40 Hz was the top end of our natural performance. This supposed limit was printed in textbooks and medical journals and accepted as a natural boundary of human experience. It was only when the brain waves of Tibetan monks exceeded 40 Hz, while they were meditating under laboratory conditions that reliably captured measurements of their brain waves, that the scientific and medical communities had to concede that we humans are capable of more than the previously accepted limits.

With no external catalysts, no plant medicines, no chemicals, and no headsets pumping electronic sounds into their ears, the meditating monks demonstrated that they were able to alter the frequency of their brain waves to double the previously accepted limit, achieving an astounding brain state of 80 Hz. A new brain state had to be defined, complete with a new name that would reflect what the monks had accomplished. Thus, the *gamma* brain state was identified that encompasses the frequency range between 40 and 80 Hz.

When the results were verified, even skeptics had to admit that the monks' ability to transcend what was believed to be an immutable human limit wasn't simply a miscalculation. This development forced the scientific community to accept that they had misjudged the capacity of the human brain, at least for some people.

Reflecting upon how they were able to achieve the gamma brain state, the Tibetan monks described how, by using yet another kind of meditation, they could exceed the newly revealed limits for brain states, pushing the capacity of the human brain even beyond the gamma brain state into new and uncharted territory.

Using nothing more than a shift of breath, focus, and awareness, the monks were able to extend the initially defined limits of the newly discovered gamma brain state up to 100 cycles per second, or 100 Hz. Their meditation techniques then doubled this already astonishing accomplishment to achieve a stunning 200 Hz, reaching another new realm of brain activity, one that's now known as the *hypergamma* brain state.

At the time of this writing, the textbooks state that the hypergamma brain state is the maximum possible frequency for the human brain. As we'll discover in later chapters, in all probability, this limit will be exceeded yet again as we learn to adapt our brains to the challenges of life in ways that we are only now becoming aware might be possible.

The point here is that what was once accepted as an immutable human limit was shown to be only a perceived limitation. As the conditions of breath and focus were shifted and refined, the previous limits gave way to higher frequencies of electrical activity in the brain, and states of consciousness previously thought to be impossible for humans were achieved.

This is an example of what divinity is all about. It's our ability to *transcend* limits of our humanness that we've accepted, or imposed upon ourselves, in the past.

## THE SECRET OF ALL SECRETS

Within our divinity is held the secret of all secrets. It's through the expression of our divine nature that we awaken an extraordinary force that is so rare, so beautiful, and so powerful that there are beings and organizations in the world that will go to any extreme to keep its presence hidden from us.

**PURE HUMAN TRUTH 7:** The battle between good and evil is ultimately the battle for human divinity.

Mysterious societies were established over the centuries in an effort to confine the knowledge of this power—our human divinity—to a few selected individuals. Nations have gone to war with nations in order to distract us from the part of ourselves that holds the secret. Armies have destroyed armies, cities have been leveled, banking systems have been collapsed, diseases have been unleashed, societies have been destroyed, climates have been engineered, and lies about our origin, our relationship to the world, to God, and our ultimate destiny have been fabricated, all in an effort to distract us from the truth of the extraordinary power within us.

The efforts to distract us continue today. As you read these pages, they're playing out in the plain sight of current world events.

Not all the events designed to keep us from recognizing our power come from the world around us, however. In many instances, the circumstances that distract us from seeing and embracing our own power come from within us. Knowingly, and sometimes on a subconscious level, we'll create situations of adversity in our lives to hide the truth of our own sacred power from ourselves. We'll create difficult relationships, financial hardship, health crises, career failures, and even life-threatening situations to occupy our awareness and distract us from revealing the game-changing secret that we each harbor within our existence.

In what may be one of life's greatest ironies, it is these very distractions and hardships that often become the lessons that reveal themselves to be our greatest teachers. They serve as catalysts for awakening the sacred power that was entrusted to us so very long ago.

Through the biology of our natural cells, specialized neurons, and DNA, we are each linked to something that exists beyond our physical body. We are fine-tuned to a timeless part of ourselves that is the source of our joy, imagination, creativity, innovation, and healing. While this essence is sometimes taken for granted in casual conversations, it's only recently that modern science has acknowledged its existence and the role it plays in our lives.

A 2004 paper printed in the scientific magazine *Journal of Alternative and Complementary Medicine* describes our relationship to this power, stating: "There is compelling evidence to suggest that the physical heart is coupled to a field of information that is not bound by the classical limits of time and space."[3] In modern language, this open-access, peer-reviewed paper describes how the biology of the human heart functions as a direct conduit—a spiritual hotline—to a field of information, knowledge, and abilities that is not limited by, or bound to, the laws of physics as we understand them today.

It's our ability to access this field through the conduit of our natural bodies—our pure humanness—that gives us the extraordinary potentials of our divinity. And this is precisely why the proposed merger of our natural bodies with the technology of computer chips, AI, various gene therapies, and nanoparticles poses such a threat to our existence.

Accepting digital technology into our bodies blocks this sacred connection and prevents us from accessing our greatest potentials by veiling the power of our divinity.

## RECOGNIZING DIVINITY

Divinity is more than our thoughts. Divinity is more than what we believe. Our divinity is more than our conscious mind. It's beyond our subconscious mind. Divinity is an expression of a part of us that is known as our *superconscious*. This is the part of us that is lasting, ancient, and timeless.

Our superconscious is the source of our direct knowing. It's the key to our deep intuition, creative imagination, and expression. It's where our self-acceptance and self-love begin, and because of this acceptance and love, it's also where our deepest states of healing begin.

**PURE HUMAN TRUTH 8:** Divinity is the part of us that is ancient and timeless, where our direct knowing, imagination, creativity, self-acceptance, and self-healing begin.

An often-used yet anonymous quote that helps to clarify the distinction between subtle states of consciousness is: "The superconscious mind is soul, source, love, the authentic you. The subconscious mind is what you are. And the conscious mind is what you do."[4] We see expressions of divinity in the world around us each and every day of our lives. Sometimes those expressions appear in unexpected ways.

## EXAMPLES OF DIVINITY

In 2018, I had the opportunity to attend the Grammy Awards celebrations, which were held in New York City that year. Amid the performances and supporting events taking place following the award ceremony, there were opportunities to meet with the singers, songwriters, and musicians who were being celebrated.

I took advantage of the access to such amazing talent, and in the natural course of conversation, made a point of asking each person I met the same question: "When you created that amazing song or you wrote those powerful lyrics," I would ask, "where did the music and the words come from?"

Without exception, each artist responded with the same answer. Every single one clarified that the words and music did not come *from* them. Rather, they came *through* them.

I had similar conversations with scientists and engineers when I was working for corporations. When a computer scientist colleague would create a beautiful and efficient piece of software that made our work lives simpler, or a mathematician would create an eloquent equation to solve a problem that our teams were facing, they would often say that they'd simply surrendered to the process. That they stepped aside and let inspiration flow through them.

The same is true for painters, sculptors, and writers with whom I've spoken about their internal creative processes over the years.

In each of these examples, it's clear that the inspiration for the greatest expressions of our creativity begins in a place that is more than our conscious thoughts. It comes from something beyond us—our divinity—which makes itself known as it is expressed through imagination, vision, and innovation. And this is precisely the reason that there is a battle for our divinity.

As powerful as our musical, visual, and engineering creations may be, our divinity is more than a visual inspiration or the solving of a mathematical equation. Divinity is the essence of our true nature. *It's the expressions of our divinity that free us from fear.* And fear is perhaps the most valuable commodity in existence when it comes to the ancient battle between good and evil and the attempts to control individuals, families, communities, societies, and even nations.

## DIVINITY EQUALS FREEDOM

Divinity frees us from the fear that keeps us feeling small, insignificant, and powerless. It allows us to become the greatest expression of our humanness and to live the best, highest, and mightiest versions of ourselves.

When we do, we find the freedom to love fearlessly. We find the knowledge that tells us we have choices in our lives. We find the wisdom to act upon those choices so we may accept our greatest gifts and extraordinary potentials. We discover the strength to follow through with what we've chosen to achieve the best possible version of ourselves.

**PURE HUMAN TRUTH 9:** Expressing our divinity frees us from the fear that keeps us feeling small, insignificant, and powerless, allowing us to triumph over life's challenges.

In the untethered expression of our divinity, we become less vulnerable to the fear projected upon us by others, and ultimately less susceptible to the power and control of others. Whether it's fear within a domestic relationship that prevents us from making healthy life choices, fear within the halls of a corporate power structure about making damaging decisions, or fear about the competing visions and agendas of our political leaders and how we will be affected by them, the expression of our divinity is what empowers us to live in joy, freedom, and sovereignty and to change the world when we see it needs changing.

The key to triumphing in the battle for our freedom is to think, and live, beyond the old ideas of what it means to "win" and "lose." To see life in such terms is precisely what keeps us stuck, struggling, and locked into perpetual conflict. The way we triumph is to elevate ourselves beyond the win/lose thinking by living our divinity in our everyday lives.

It's by loving fearlessly, forgiving without expectation, honoring our bodies with the highest form of nutrition available, and learning to trust the innate intelligence of our immune system (and other bodily systems) that we *triumph over* the oppression we see around us rather than attempting to win a battle by fighting *against* it. By celebrating our divinity and living as the best version of ourselves, we transcend the polarity of the good and evil to express what it means to be a pure human.

In two brief phrases, 19th-century Indian philosopher Hazrat Inayat Khan eloquently described the relationship between our humanness and our divinity. He began by first identifying the reality of our perceived human limits, stating: "Humanity, divine limitation."[5] In these three words, he conveys our near-universal sense that we humans are limited expressions of something less than the divinity that is possible.

The second part of Kahn's statement brilliantly resolves the mystery of this relationship, stating: "Divinity, human perfection."[6] Here, he reminds us that by living the truth of our essence, we lift ourselves into the realm of godlike qualities that reveal the entirety of what it means to be human.

Through these six brief words, our divinity is described as being more than simply an optional part of us that we may choose to express at selected times in our lives. Rather, our divinity *is* us. It's the whole of us. It's what makes us complete.

It's through the access to our divine capacities, such as imagination, intuition, empathy, forgiveness, and compassion, that we open the door to the fullest expression of our physical and spiritual healing and what it means to be human.

## DIVINITY BEGINS WITH OUR STORY

We embrace our divinity through the way we think of ourselves—our story. For some people, the word *story* conveys the sense of a casual or philosophical conversation that is separate from the reality of their everyday lives. When we look deeper, however, we discover that nothing could be further from the truth. It's clear that our personal story is beyond just an interesting sidebar in our lives.

The way we think of ourselves is the essential part of our lives that creates the path to our wholeness and healing. It's what completes us. We live our lives, choose our partners, heal our bodies, identify our spiritual values, and claim our politics based upon our learned sense of self-worth and through our lived relationships with people and life.

**PURE HUMAN TRUTH 10:** Awakening your divinity begins with the way you think of yourself—your story.

The consequences of the way we think of ourselves drive our actions and the choices we make in each moment of every day in our lives. From what we eat and how we arrange our day, to our relationship with money, love, death, and health, it's our individual story about who we are that determines how we respond to the challenges of life, to other people, and ultimately, to our relationship with God.

When we consider the role of our self-view or personal story from these perspectives, it's clear that our story provides the foundation for all that we do, all that we dream, all that we aspire to achieve, and the most cherished and heartfelt experiences of our lives.

With the significance of our story in mind, the power of the way we think of ourselves is clear: If we change our personal story, we change our lives. If we change our collective story, we change the world. And it's precisely for this reason that we need to ask ourselves, each and every day, a single simple question: What is the story that I tell myself, about myself, and believe?

## RETHINKING OUR TRADITIONAL STORY

As I pushed my shopping cart along the fresh-produce aisle during the first week of the new year, a woman thought she recognized me from a recent series she'd seen on YouTube. She approached me cautiously and in an uncertain voice asked, "Are you Gregg Braden?" I've learned that this can sometimes be a tricky question to answer, depending upon how the person that's asking feels about the content I shared in those or other videos.

I replied with a question of my own. "I don't know," I said with a smile. "If I am Gregg Braden, is that a good thing? Or a bad thing?"

The woman caught on to my hesitation quickly and laughed, sharing that she just wanted to thank me for the programs and to let me know how much she appreciated the clarity that they brought to the nightly discussions she shared with her family at the dinner table.

She summed up her feelings about the chaos of the present-day world by saying that something seemed "off" and the world just didn't feel right. "Something's wrong," she said. "Things are moving too fast. We're not choosing the changes in our world, and they're not coming *from us,* they're happening *to us*. We need to slow down, take a deep breath, and get back to the basics in life. We need to live more simply and remember our relationship to nature."

Everything the woman said spoke to me on a deep level. She was telling me that we need to rethink the way we're living and the way we're thinking. In her own way, she was saying that we need to change the stories we tell ourselves about ourselves.

Our stories come from many places. They're the composite of ancient traditions and perspectives that we're indoctrinated to accept throughout the various ages and developmental stages of our lives. The foundation for our personal story begins at a young age with our immediate family. This is when we pick up on the way our caregivers deal with life. Among other things, we learn from our parents who the "good" people in the world are and how to recognize the "bad" ones. Through us watching their interactions, we learn how to enjoy people that we like, and how to deal with friends and neighbors that we disagree with.

The blueprint for the way we think of ourselves and what we believe we're capable of further deepens through our experiences in the classroom and through the perspectives we read in our textbooks and are immersed in via the community support of our religious upbringing. The society we're born into plays a powerful role in forming our story too. Our friends, neighbors, culture, and personal experiences all add to, refine, and solidify the way we think of ourselves.

But our personal story goes even deeper. Through the work of scientists like neuroscientist and pharmacologist Candace Pert, who wrote the landmark book *Molecules of Emotion,* we now know that our emotions are constantly creating chemicals in the body that represent our perceptions of our life experiences. These chemicals are in the form of molecules called *neuropeptides*.

In simple terms, neuropeptides are chemical messages that our bodies produce from the way we *feel* about our experiences. I'm emphasizing the word *feel* here because this is the key to understanding the role that neuropeptides can play in our emotional, as well as physical, healing. While the quality of our experiences themselves is certainly important, the neuropeptides in our bodies are linked less to the facts of our experiences and more to the way we *feel* about them, and the significance we give to them.

The conditioning that forms the way we respond to the world actually begins even *before* we arrive in this world. It begins while we're in our mother's womb. From the moment that we're conceived, we are intimately linked to the experiences, emotions, and chemistry that our mother creates as she carries us in her body.

## OUR PERSONAL STORY BEGAN IN THE WOMB

The neuropeptides that our mother produced from her experiences of life while she was pregnant with us coursed through her blood, organs, and tissues and through ours as well. Her chemistry primed us for what we needed in the world we were born into. Nature's assumption is that, at least initially, we would be in the same environment as our mother, and therefore we'd face the same challenges in the early years of our lives that she faced when she conceived us. In our mother's womb, our story began, and it programmed us either for healing and regeneration or for the "fight or flight" that we needed to deal with the world we'd soon emerge into.

For example, if our mother felt safe, nurtured, and loved in her world, then her sense of safety and well-being signaled her heart and her brain to produce a healing and rejuvenating chemistry that reflected her feelings.

If, on the other hand, our mother was in an environment where she felt threatened and she experienced the anxiety, stress, and fear of not feeling safe, then her body produced the chemistry that reflected those perceptions.

In either instance, the chemistry of our mother's blood flowed into us in the womb. It influenced us in ways that ranged from our body weight and size to our brain size and cognitive abilities. In recent years, this relationship has been very well documented in refugee camps that were developed to accommodate families that had been displaced as a result of disasters and war.

Some of the most extensive of these studies are the result of the civil war in Syria that began in 2011. Through the humanitarian tragedy that occurred in the years that followed, approximately 5.6 million people fled the country for safety. Nearly 50 percent of them were children. At the time of this writing, the crisis persists. It has continued for so long that an entire generation of children has now been conceived, born into, and knows only the harsh, often threatening conditions of life in a refugee camp.

A 2021 study of Syrian families who relocated to Turkey clearly documents the relationship between the environment of the refugee camps and the cognitive development of both parents who present with war-related mental health problems, like post-traumatic stress (PTS), and the emotional processing abilities of their children.

The study found that "high levels of maternal PTS negatively impact children's emotional processing development."[7] Perhaps not surprisingly, while both the father and mother may have been suffering from the effects of the war, it was the mother's levels of PTS during her pregnancy that had the greatest impact on the children in this study.

A second study published by the humanitarian organization World Vision helps us to understand the mechanism of the stress–cognitive development relationship in children. The report concludes: "Neglect and under-stimulation of children affected by conflict can lead to severe impairments in the cognitive, physical and psychosocial development of the child, creating a lasting legacy of war. This can lead to emotional, cognitive, and behavioural disorders, anxiety and depression, emotional and interpersonal difficulties, and significant learning difficulties."[8]

The tragic examples from the refugee camps leave little doubt as to the importance of our prebirth experiences, as well as our early childhood, when it comes to the way we think of ourselves.

While we have no control over the world our parents find themselves in before our birth, the good news is that unhealthy responses that we may have inherited at birth and during our childhood are not "set in stone." Using epigenetic triggers that I'll describe throughout this book, they can be modified and revised to create healthy stories, present-day healing, and healthy responses to events in our lives.

This capacity for modification is important because it means that we do not have to be defined today by difficult circumstances in the past. That is, unless we choose to be. Nature empowers us with the ability to change the significance that we give to the betrayal, trauma, and loss of our past. Each time we do so, we change our story—which changes what's possible for us.

It's for this reason that the latest discoveries about the origin of our species play such a powerful role in the way we've been taught to think of ourselves.

## THE STORY OF OUR PLACE IN THE UNIVERSE

While our personal story begins in the mystery of our mother's womb, our collective story begins in the womb of creation with the mysterious origin of the universe. Conventional science tells us that we're the product of a dead universe, made of sterile and inert material, that began approximately 13.8 billion years ago with the event known as the Big Bang. It was following this massive, primal release of energy that we're told a series of additional events occurred that are so unlikely, yet so unbelievably perfect, that they border on the miraculous. But new discoveries are now telling a very different story.

While it's still a mystery as to precisely what existed before the Big Bang, and why the Big Bang even happened, current computer models that mimic the beginning of the universe show that the discharge of energy during this occurrence did something almost unthinkable in today's terms. One of the reasons it's hard to imagine is because of the way the energy itself was released.

When we think of a conventional explosion, we typically imagine the kind we see in Hollywood action movies, made with dynamite. The blast begins with a boom, a sudden flash of light, and a burst of energy that originates in one central place and then expands outward in all directions, like a Fourth of July firecracker popping in slow motion.

When it comes to the Big Bang, however, a very different kind of explosion took place. Rather than happening at a specific place *in* space, the birth of the universe was an explosion *of* space itself. The Big Bang literally created the space that it would expand into and now occupies. And the process is not finished yet. If the sensors on our satellites are correct, the expansion of energy and the resulting creation of new space are ongoing and continue today.

While the origin of the Big Bang remains a mystery, scientists agree that it was in those first few fractions of a second following the initial burst of energy that the properties of matter, space, and time that we study through physics were established. It's also during this time that all energy that would ultimately condense into matter was still unified in what physicists call the *singularity*: a point in the history of the universe when everything was physically and energetically connected.

Although modern mathematics breaks down when it comes to describing the infinite temperatures and volume of the singularity, cosmic background radiation data gives us a sense of the conditions during this time. When the young universe was only 10 million trillion trillion trillionths of a second old, the mind-boggling temperatures were in the range of $10^{32}$ Kelvin, or 180 million trillion trillion degrees Fahrenheit. While these measurements may seem meaningless to some people, and still leave the scientists who have revealed them in awe, they convey a sense of just how extremely hot the universe was in just the first few fractions of a second of its existence.

Then, due to reasons that are still not fully understood, an unlikely and seemingly miraculous sequence of events was set into motion as follows.

- The energy of the young universe began to expand, cool, and condense *in just the right way*, at just the right time, to form the first atoms of matter.
- Those atoms then combined *in just the right way* to form the first simple elements listed on the periodic table.

- Those elements then began to transition *in just the right way* to form massive clouds of gas.
- The clouds eventually coalesced and condensed *in just the right way* to form the stars and planets that we can see in the sky today.

## THE RIGHT PLACE AT THE RIGHT TIME

Following this already unlikely series of events, our planet then formed in just the right way to create the conditions to produce life, and eventually, to produce humankind. Scientists often refer to this series of unbelievably lucky, habitable conditions as Earth's *Goldilocks zone*—the optimum temperature, atmosphere, and climate that support life as we know it.[9]

In 1961, Princeton University physicist Robert H. Dicke recognized that the number of conditions needed for our planet and life on our planet to exist as they do are just too many, too complex, and too precisely "tuned" to one another to be purely the result of coincidence.[10] He also realized that, if even one of these parameters were to vary slightly above or below the numbers that define them in our world today, then Earth life as we know it could not have formed as it has, and could not exist as it does.

For example, if the universe were approximately 10 times younger than it is today, not enough time would have passed to build the density of elements needed to create small, rocky Earth-size planets that are optimum to support life. If the universe were 10 times older, many stars, including our sun, would have progressed so far into their life cycles that they would be only dense remnants of their former states, known as *white dwarfs*.

While the number of these Goldilocks conditions varies based upon how detailed the reports are made to be, I'll identify seven of them here just to give you a sense of what they are and just how finely tuned the universe and our place in it really are.

**Goldilocks condition 1: A perfect distance from the sun.** Water is essential for life. Earth's orbit places it in exactly the region necessary for water to exist and be accessible to life in a liquid state.

**Goldilocks condition 2: A perfect magnetic field.** Our planet is surrounded by a magnetic field that protects life from cosmic radiation. This field is the result of Earth having a molten inner core and its rotating layers (outer core, mantle, and crust) moving around that inner core to produce magnetism.

**Goldilocks condition 3: A perfect atmosphere.** Our planet has just the right mix of life-affirming gases (including carbon dioxide, nitrogen, and oxygen), and the perfect density of those gases, to support and be accessible to life.

**Goldilocks condition 4: A perfect amount of rocks.** Rather than the dense and liquified gases of some large planets in our solar system, such as Jupiter and Neptune, the size of Earth and the composition of Earth's crust statistically offer the greatest opportunity for life and for a diversity of life to arise.

**Goldilocks condition 5: A perfect temperature.** The average temperature of our planet is 59 degrees Fahrenheit / 15 degrees Celsius. Unlike the temperatures for other planets in our solar system, and beyond, this relatively constant temperature allows for water to remain in the liquid phase and for life to exist optimally.

**Goldilocks condition 6: A stable sun.** Scientists estimate that approximately 85 percent of the stars in our galaxy are binary star systems. This means that there are two stars orbiting each other. The impact that two such massive solar bodies would have on the gravity of a planet like ours would make life on Earth difficult. With the exception of the large-magnitude solar flares that have occasionally happened in the past, our sun is a relatively stable star.

**Goldilocks condition 7: A perfect amount of water.** Earth's water is believed to have resulted from collisions with icy comets between 3 and 4.5 billion years ago. The result is that we have more water than land mass and abundant water to sustain life.

If these conditions, and others, are truly the result of purely random events that occurred purely by chance, then our world is the product of the most unbelievably lucky physics imaginable.

**PURE HUMAN TRUTH 11:** We're more than the result of random processes. It is statistically beyond chance that the seven Goldilocks conditions that make our world and our lives possible are the result of "lucky" physics.

The point of this discussion is that when it comes to our story, we're told that we're the product of a dead universe that was formed by a lucky series of events and void of consciousness, intelligence, and meaning. The consequences of believing this uninspiring story are reflected in the casual way we are led to think about life in general, our lives specifically, and our relationship to the world around us, to one another, and to the choices we make each and every day.

## OUR LIVES REFLECT OUR STORIES

While the details of the events that scientists use to explain the cosmos may seem academic, and for some people even philosophical, the implications of our cosmological story reverberate to the core of our society today. The consequences of believing that we are the product of a dead universe remove any reason for reverence and respect when it comes to life, nature, and natural resources. And it's that lack of reverence that becomes the implied license to rewrite the codes of life, engineer our bodies, and exploit the natural world.

We see this thinking reflected in the way modern society has viewed Earth's resources as natural assets to be taken advantage of, rather than as a relationship to be honored and protected. One of the leading minds defining how the new scientific paradigm of a dead universe affects us in our daily lives is author, educator, and consultant Duane Elgin. He shows us that the way we think of the universe, and our place in it, is at the very foundation of the way we live our lives and solve our problems, especially when it comes to how we treat one another.

In Elgin's words, we relate to our belief that we're in a nonliving universe "by taking advantage of that which is dead on behalf of the living. Consumerism and exploitation are natural outcomes of a dead universe perspective."[11] With a few exceptions, Elgin's statement describes the way that much of humankind has lived in the past and continues to live today. The problem with this mindset is that, ultimately, it has led to the depletion of natural resources, unsustainable forms of food and mineral production, and the types of conflict over scarce resources that are at the root of so much suffering today.

The good news is that recent discoveries are overturning the old cosmological story and potentially leading us to hold a radically different view of the universe and our place in it.

## THE UNIVERSE IS ALIVE

New discoveries are now supporting the perspective of our ancient ancestors when it comes to the story of our universe. A growing body of evidence suggests that the universe is far from dead and inert as previously believed. In light of the new evidence, it appears to be not only alive, but also, as we'll discover later in this chapter, both conscious and intelligent.

The idea of a conscious universe has led to an emerging scientific theory known as *panpsychism*. The word itself comes from two Greek roots: *pan*, meaning "all," and *psyche*, meaning "mind" or "soul." The essence of panpsychism is that, in addition to being associated with the minds of living beings, consciousness may be inherent in things that are not what most of us would typically think of as being "alive."

This thinking accepts consciousness as a force that may, in fact, be a universal phenomenon that extends throughout the universe to include nonliving systems. So the language that some scientists are now using to describe the universe literally means *all mind*. And this is precisely the sense that we get when we take a deeper look at the evidence for panpsychism.

The growing popularity of this theory as a respected scientific field of study is stated in a paper by Gregory Matloff, a physicist teaching at New York City College of Technology, CUNY. In the abstract for his research paper "Panpsychism as an Observational Science," Matloff states: "The work of two separate researchers who are contemplating methods of communicating with stellar-level consciousness indicates that experimental astro-panpsychism may be possible as well as observational astro-panpsychism. It is becoming evident that panpsychism may be moving from the realm of metaphysics to the domain of observational astrophysics."[12]

In an earlier research essay published in *Journal of Consciousness Exploration and Research,* Matloff clarifies the implications of his theory and what panpsychism means for the study of cosmology. "According to panpsychism, consciousness is built into the fabric of the universe."[13]

Freeman Dyson, one of the most significant mathematical physicists of the late 20th and early 21st centuries, clearly favors panpsychism. In his acceptance address for the prestigious Templeton Prize in 2000, Dyson stated, "Mind seems to play a role in at least three levels in the universe—the quantum level of elementary particles, the human level, and the cosmological level at which universal laws seem fine-tuned to allow the emergence of life."[14]

One of the reasons for the shift from thinking of the universe as a dead system to the living and intelligent system that Matloff describes is due to observations of stars and star systems that seem to show them responding to cosmic disturbances in ways that reflect how living systems respond to changes in their environment.

For example, the images received from both ground and space-based imaging of distant galaxies have revealed previously unrecognized and mysterious jets of gamma rays emitting from the centers of numerous galaxies.[15] The energy of these rays acts to shift the location of the star systems over time in ways that appear to move them to safer locations during violent cosmic disturbances. Through his published research, Matloff suggests the possibility that these jets may be showing that some stars *actually respond* to galactic shifts in ways that appear to be intelligent and cannot be attributed to chance or coincidence.[16]

**PURE HUMAN TRUTH 12:** Our universe appears to be alive, conscious, and intelligent.

Matloff also describes the experiments that need to happen to either disprove or confirm his theories. If they are confirmed, Matloff's theories of the galactic jets may be some of the best evidence to date for the "mind" that Dyson described at the cosmological level. As the technology becomes available in the years to come, Matloff will have the opportunity to test his hypothesis. And if the experiments bear out what he's presented in his writing, we'll see the implications of the living universe ripple beyond classrooms and textbooks into industry, society, and our everyday lives.

## IN A LIVING UNIVERSE, LIFE MAKES SENSE

In a universe that's alive, it makes sense that life would appear often and be expressed in many and varied forms. It makes sense because life itself is the force that's driving the system. To discover that we exist as living beings within the context of an even larger living system implies that our individual lives have a deeper meaning. We're about something more than the lucky biology of simply being born, enjoying a few great years of love, chocolate, and rainbows on Earth, and then dying. It implies that in some way, underlying everything we know and see in our everyday world, our lives have purpose.

And this is precisely where we find ourselves as a society. We're at the crossroads of two ways of thinking about ourselves within the context of the universe we live in. The living universe that Elgin, Matloff, Dyson, and others describe offers us the big picture of life having a purpose from the top down—from the macro scale of the universe itself, as a living entity, to the micro scale, where the living cells and particles that make up our bodies are an expression of the life theme that permeates creation.

The discoveries that I will describe in Chapter Two offer the evidence from the bottom up—from the micro world of mutated DNA yielding more complex expressions of life and the capabilities of our lives to the macro context of a living universe. When we consider that the universe is something alive and that we're part of that aliveness, it changes everything about our story.

Duane Elgin's words offer a beautiful sense of this perspective. He says, "In a living universe our physical existence is permeated and sustained by an aliveness that is inseparable from the larger universe. Seeing ourselves as part of the unbroken fabric of creation awakens our sense of connection with, and compassion for, the totality of life. We recognize our bodies as precious, biodegradable vehicles for acquiring ever-deepening experiences of aliveness."[17]

The existence of a living universe tells us that we are part of the world around us and not separate from it; that our personal aliveness is part of an even greater aliveness. And as the very goal of life in the universe is to grow, change, and perpetuate itself, these are precisely the qualities that we should strive to embrace throughout the course of our time in this world.

Through each experience we face in life, we learn to know ourselves better as individuals and as a species, as life expressing within a container of aliveness.

This is the very definition of a living universe and our role in it. Our lives and lifetimes are our way of infusing the essence of our unique experience into an already living and extremely diverse entity. Science fiction writer Ray Bradbury sums this up perfectly, stating: "We are the miracle of force and matter making itself over into imagination and will. Incredible. The life force experimenting with forms. You for one. Me for another. The universe has shouted itself alive. We are one of the shouts."[18]

Within the limits that science has placed upon itself today, there is no direct way to know the purpose of life with certainty. Indirectly, however, the answer to the question of life's purpose may be hidden in plain sight. We may discover that the very existence of our advanced capabilities, like intuition and empathy—qualities of our divinity—holds the key to solving this mystery.

The beauty of Bradbury's statement is that it transcends formulas, algorithms, and logic. It's a purely intuitive answer to a serious scientific question. It's also a perfect example of how advances in modern science have carried us to the edge of what science can tell us with certainty. There's a place—an unspoken boundary—where the nuts and bolts of scientific explanation fail when it comes to describing the essence of life. They fail because we're more than a random collection of cells, flesh, and bones. There's a quality to human life that simply cannot be defined in purely scientific terms, at least not as we know science today. And it's that quality that can lead us to comprehend the deepest truths of our existence.

## DARWIN'S THEORY WORKS, UNTIL IT DOESN'T

Similar to the way we've been led to think of the universe as the product of mysterious and lucky physics, conventional science has led us to think of ourselves as the result of equally mysterious and lucky biology. When it comes to the origin of our species and life on Earth in general, the prevailing scientific theory suggests that we are the product of slow and gradual changes that occurred over a long period of time—Charles Darwin's idea of evolution. I want to be very clear about what I'm saying with respect to evolution in general and what it means for our origins.

As a geologist, I fully believe in, and support, the discovery of evolution as the mechanism that led to the emergence of many forms of life, including early primates. During my undergraduate geology fieldwork, I searched for, and recovered, the fossilized remains of many forms of life that clearly support evolutionary theory, including plants, insects, and marine life. Darwin's theory breaks down when it comes to us, however, and is not supported by the physical evidence we've uncovered.

It's already been well established that we modern humans are not descendants of the Neanderthals, for example, as previously believed.

The fact that our most recent genetic science shows that our forebears interbred with them is evidence that we came from separate lineages.

It has also been established that the genetic mutations that are responsible for our humanness, including our ability to embrace empathy, compassion, honesty, and moral judgments, and to do these things on demand, are made possible by the mysterious genetic mutations that gave us the ability to do so. Though the mutations themselves could be the subject of an entire book, for convenience I'll briefly summarize one of the mutations that tell us why evolution is not our collective story.

## THE ANCIENT STORY OF MODERN HUMANS

When I was in school during the 1950s, 1960s, and 1970s, the thinking was that, in addition to the familiar precursors to modern humans, such as Neanderthals, *Australopithecus* (the famous Lucy), and *Homo habilis* (known as the "handy man"), there was another member of the evolutionary family tree who was a close ancestor as well. At the time this ancestor was called Cro-Magnon. Since that time, however, a new name (one that makes more sense) has replaced the familiar name of the past. The new name is *anatomically modern human,* which is typically referred to by the acronym AMH.

The scientific community generally agrees that AMHs appeared on Earth about 200,000 years before the present. But unlike other forms of life that became extinct long ago, including human relatives such as Neanderthals, AMHs never disappeared by becoming extinct. They have a continuous lineage of DNA that can be traced into the present, and they remain on Earth today, visible on every continent of the world. They populate the largest cities in every country and are the most obvious form of life dominating our world today. They are us. We are them.

For all intents and purposes, we are the AMHs. The same technology that tells us we did *not* descend from Neanderthals now reveals that we *are* the anatomically modern humans who appeared mysteriously 10,000 generations ago.

**PURE HUMAN TRUTH 13:** The first of our kind appeared on Earth approximately 200,000 years ago, we're still here, and our DNA blueprint hasn't changed.

The undeniable fact of our existence, and the mystery of our origin, is a problem for Darwin's theory when it comes to the evolutionary tree of life. Rather than evolving our unique capacities and attributes slowly and gradually over a long period of time, as evolutionary theory would suggest, we appeared on Earth relatively suddenly. When we did, our humanness was already fully developed and enabled. In evolutionary terms, we were ready to go and "hit the ground running," as the expression goes, when we arrived. Although ancient humans may not have behaved exactly like we do, they looked like us, they functioned like us, and they appear to have had all of the capabilities of intuition, empathy, and self-regulation that we have today.

Recent investigations into the way our genome was formed have shed light on the lingering mystery of how our species could have broken the rules of evolutionary theory and arrived as we did. The investigations reveal nuances of our existence that Darwin could not possibly have realized with the limited technological resources available to him in his time. One of those nuances has also revealed one of the greatest mysteries of our existence: the mysterious mutations that led to the existence of one of the largest chromosomes in our body.

## THE MYSTERY OF HUMAN CHROMOSOME 2

Human chromosome 2 is the second-largest chromosome in the human body. It represents about 8 percent of our genome and, depending upon the research method used to make the count, appears to consist of between 1,200 and 1,300 genes. In addition to being one of the largest portions of our genetic code, it is also one of the most mysterious. The analysis reveals that this unusually large chromosome is actually made of two smaller, preexisting chromosomes that have been fused together, end to end, at some point in the distant past in a way that makes them appear as one large and complex chromosome.[19]

In other words, at a moment of time in our ancient past, for reasons that remain uncertain as well as controversial, two separate and independent chromosomes that are still present in the DNA of our nearest primate relatives were merged into a single larger chromosome, and then tweaked to optimize the fusion and produce what is now our chromosome 2.

New DNA technology has revealed that the fusion of the chromosomes that resulted in our modern chromosome 2 is almost a certainty. We now have the technology available to us that can replicate such a fusion to reveal precisely how the preexisting chromosomes forming chromosome 2 were combined.

In the following bullets, I'll summarize the essence of this discovery in two ways. First, I'll share the scientists' own words as reported in the *Proceedings of the National Academy of Sciences* to announce the 1991 discovery itself. Second, I'll share a simpler description in my own nontechnical terms, before elaborating on why this discovery is so important to our discussion of our human story.

- **The technical explanation.** *Proceedings of the National Academy of Sciences of the United States of America* describes the origin of chromosome 2 as follows: "We conclude that the locus cloned in cosmids c8.1 and c29B is the relic of an ancient telomere-to-telomere fusion and marks the point at which two ancestral ape chromosomes fused to give rise to human chromosome 2."[20]
- **The simplified explanation.** It appears that long ago two separate chromosomes from primates (chromosomes 2A and 2B) fused into the single, larger human chromosome 2, which is one of the key chromosomes that give us our special humanness.

Many of the characteristics that make us uniquely human are only possible because of this mysterious fusion. These include our capacity for intellect, the growth and development of our brain in general, and more specifically of the largest part of our brain, the neocortex, which is associated with the way we think and act, and gives us our capacity for emotion.

While chromosome 2 contains over 1,200 genes that continue to be mapped and studied today, what we do know is that it provides the instructions for some of the most vital functions of our existence. In the table below, I'll share just a few simplified examples of these genes to give you a sense of the vital roles they play in our lives.[21]

**A Sample of Genes from Chromosome 2 and What They Do for Us**

| Human Gene | Influence in Our Body |
|---|---|
| TBR1 | Fundamental to brain development, especially the neocortex that is associated with our capacity for emotion, empathy, and neuron functions. |
| BMPR2 | Central in osteogenesis (bone tissue formation) and cell growth throughout the body. |
| MSH2 | Known as the "caretaker" gene, its chief function is tumor suppression. |
| SSB | Involved in the fetal development of organs. |

From this small sample of genes, it's clear that chromosome 2 plays a vital role in contributing to who we are, as well as what we are. In light of its significance to our identity, the question of how chromosome 2 came into existence is now more important than ever as the movement to embrace transhumanism gains momentum.

As is often the case when exploring a mystery, the answer to this question is contained within the mystery itself—it's found within the chromosome itself. Similar to the way the digital blockchain of a Bitcoin contains a transparent record of every transaction that precedes the current entries, chromosome 2 has also preserved a record of each of the ancestral mutations that make us as we are.

This record for our species as a whole is intact and transparent, and we now have the technology to read the record and reveal what it says. But as happens so often when we attempt to solve a mystery, although we may answer the original question that has been asked, the answer opens the door to an even deeper mystery. Solving the mystery of chromosome 2 was no exception.

## AN UNNATURAL FUSION

The fusion of chromosomes that produced HC2 happened in a way that scientists say is improbable at the very least, and almost impossible under natural circumstances. The reason they say so is that this fusion happened between special segments of those chromosomes called *telomeres*. Telomeres are nonvital sequences of DNA located only on the ends of a chromosome. They are important because they prevent the loss of vital genetic information whenever the cell divides.

Cell division is a traumatic experience for the chromosomes. As the genetic material is pulled apart—divided—to make the new cell, the ends of the DNA are fragmented, and the information they contain is lost during the process. Nature's remedy to preserve the essential portion of the DNA holding the instructions for the new cell is to create a buffer on the tip of the chromosome that takes the "hit" of the cell division.

I'm describing the telomeres in some detail here because they play a vital role in the mystery of our fused chromosome 2.

It's not unusual to see chromosomes break in response to environmental factors, such as exposure to high doses of radiation, overexposure to intense ultraviolet (UV) light, or in response to taking recreational drugs, such as LSD. When this type of breakage occurs, the fragments of the broken chromosome typically fuse with other fragments of chromosomes that have broken in a similar way. The important thing to note here is that this type of fusion happens at the point where the chromosome breaks, with the telomeres remaining intact at the ends of chromosomes.

This is not what happened with chromosome 2. Chromosome 2 was not the product of this kind of fusion. In fact, there appears to have been no breakage at all. Instead, two fully intact chromosomes were fused together, end to end, *telomere to telomere*, in a way that makes little sense under natural conditions. And this is the "smoking gun" revealing a previously unthinkable possibility for chromosome 2.

The fusion point of the two preexisting chromosomes that produced our chromosome 2 is clearly visible in the middle of the chromosome. In addition to the telomeres that naturally remain on both ends of this chromosome, the telomeres that were once on the ends of the stand-alone chromosomes (2A and 2B) are now fused together near the middle of human chromosome 2.

As extraordinary as this is, on its own, the mystery of chromosome 2 doesn't end with this unusual fusion. There's more!

In the language of the researchers describing this discovery, the fusion was either "accompanied or followed by inactivation or elimination of one of the ancestral centromeres, as well as by events which stabilize the fusion point."[22]

While this language is admittedly complex, the message it communicates is clear and simple. The study is telling us that during or immediately after the fusion that produced chromosome 2, the overlapping functions from what were originally two separate chromosomes *were either adjusted, turned off, or removed altogether* to make the resulting new chromosome more efficient.

Both the fusion and the mutations of chromosome 2 that followed were dated and found to have occurred 200,000 years ago, precisely at the time our ancestors emerged as anatomically modern humans. The odds of these critical processes happening by chance are staggeringly small.

## BEYOND IMPOSSIBLE ODDS

Scientists generally accept that when the odds of something occurring are $10^{400}$, the chances for that "something" happening are so small it is considered an impossibility. The odds for the mutations in chromosome 2, and others, occurring quickly, with the precision and result they produced, are considered to be $10^{600}$. If $10^{400}$ yields impossible odds, then $10^{600}$ means that the odds are more than impossible. It is more than impossible that chromosome 2 is the result of random processes and lucky biology. And this is where modern science is stuck.

The reason? Biology is based upon the paradigm of random mutations occurring slowly and gradually over long periods of time. It's not well equipped for the implications that have resulted from the chromosome 2 studies. Because the second-largest chromosome in our genome, the one that is responsible for much of our humanness, cannot be explained by evolutionary processes, at least not by those that we know of today, the implication is that there is an intentionality underlying our existence as humans. And intentionality is typically associated with purpose.

For the scientific community to acknowledge where the new human story is leading represents nothing less than a paradigm-shattering shift in perspective. And that's precisely where the story of our origin as a species is heading.

**PURE HUMAN TRUTH 14:** It is mathematically impossible that the mutations responsible for our most cherished human qualities, including empathy and intuition, are the result of random processes and "lucky" biology.

## WHO OR WHAT IS RESPONSIBLE FOR THE CREATION OF OUR CHROMOSOME 2?

If human chromosome 2 is the result of something beyond the known process of natural selection that Darwin described, then the obvious question is, "What was it?" What, or who, is responsible for the mutations in our genome that give us our humanness?

The honest answer to this question is a source of frustration to the scientific community studying the phenomenon. The truth is that scientists simply don't know.

Even now, in the early years of the 21st century, more than a quarter of a century after the fusion was initially recognized, scientists still cannot say with absolute certainty why primate DNA was fused in a manner that led to the emergence of anatomically modern humans. What we can say with certainty, however, is that the DNA that makes us who we are is *not* the result of the process of evolution that Darwin described.

And this is the point: we will never know the deep truth of our mysterious origins until we allow ourselves to follow the evidence to the story it tells.

When we can definitively determine how the ancient DNA fusion that we see in chromosome 2 happened, and how these specific pieces of the fused DNA were modified so precisely and so quickly, thousands of years earlier than any technology that could do so was known, then the solution to these mysteries will lead us directly to an explanation for why such an extraordinary event took place.

## THE NEW HUMAN STORY

We've asked science to tell us who we are. In the early years of the 21st century, science and the best scientists the world has ever known are telling us clearly the answers to our questions. The problem is that many of the same scientists who have asked the questions are now reluctant to accept the answers being revealed. The reason is that the new discoveries fly in the face of tradition, doctrine, conventional thinking, and the status quo.

We've raised our families, built our societies, chosen our policies, and built the curriculum that we teach our children based upon the belief that we're the product of a sterile and random universe. From within the echo chambers of academia, we've been indoctrinated into the story that humankind is the result of random mutations and lucky biology. The sampling of discoveries that I've shared in this chapter, and the new discoveries that are filling the pages of prestigious scientific journals on a weekly basis, tell us that we need a new story. The existing way we've been taught to think of ourselves is no longer supported by the evidence.

And while the status quo story of random mutations in a lifeless universe may have served us in the past, now we know that story is based upon the false assumptions of an obsolete way of thinking. Yet, this is precisely the thinking that now forms the barrier between us and the remarkable world of the future that in our hearts we know is possible. The first step to embracing our divinity is to gather the courage to embrace what the new discoveries are telling us about ourselves.

If we are truly the product of an intentional act of intervention that happened long ago, as the evidence now suggests, where would we look for the clues to tell us more? For millennia scholars have asked precisely this question. They've searched for the answer in the only ways they knew: looking for clues within the faded manuscripts, chiseled stone walls, and fragmented books of antiquity.

And while new archaeological discoveries have been made and new information revealed, the ancient records have yet to disclose the secret of our origin.

A new interpretation of our past suggests that the key to solving the mystery of our origin may be found within what spiritual traditions refer to as the crowning achievement of God's creation: Within us. Within our humanness. Within the cells of our bodies.

In the next chapter, we'll interpret the mysterious clues from our past through the eyes of modern science. In doing so, together, we'll solve the mystery, and reveal the message, that was coded into the building blocks of our lives—into our DNA—200,000 years ago.

CHAPTER TWO

# Who Are We?

## The Ancient Message Coded in Our DNA

God becomes as we are, that we may be as he is.

— William Blake (1757–1827),
English poet, painter, and printmaker

In what may be the greatest irony of our species, following more than 5,000 years of recorded human history, we have yet to answer with certainty the most fundamental question of our existence. Who are we?

The way we answer this seemingly simple question has implications that underpin every facet of our daily lives. The meaning we give to "Who are we?" shows up in the foods we choose to nourish our bodies and how we care for ourselves, our youngest children, and our elderly parents. It forms the lens through which we see ourselves in the world, the choices we make for our lives, and the way we build civilization itself. Our answer defines how we share vital resources like food, water, medicine, and other necessities of life. It determines when and why we go to war and what values our economies are based upon, as well as how those economies are balanced and maintained.

What we believe about ourselves, our origin, and our ultimate destiny even justifies our thinking when it comes to saving a human life, and when we choose to end a human life. Where are we to look for the answer to this most fundamental question of our existence?

# LOOK WITHIN

Is it possible that a written record, a time capsule of knowledge about our origin revealing the most sacred trust of humankind, has survived the cycles of time, awaiting discovery since the dawn of our creation? For centuries, scholars have searched for such a time capsule within the written records and oral traditions of those who have come before us. They have searched the catacombs of remote monasteries in the Himalayan mountains and Middle Eastern deserts, and scrupulously analyzed the crumbling remains of the Dead Sea scrolls that were transcribed from even earlier records, one letter at a time, over two millennia ago.

The discoveries so far, while interesting, have failed to live up to the expectations of researchers. The ancient records have yet to reveal the secret of our origin. What they have revealed, however, are clues to solving the mystery, suggesting that the records preserved in the outer world around us are incomplete. They reveal only a portion of something much greater that is to be found in the world within us.

From the original inscription above the entrance to Apollo's temple at Delphi and the strange script in the Emerald Tablets of Thoth (a controversial Hermetic text said to reveal the secrets of the universe), to mysterious passages recorded in the Kabbalah's ancient Book of Creation, there are references to a universal key that holds the long-forgotten answer reminding us of who we are, the identity of our creator, and the origin of our species. Through an eloquence and a directness that is typical of many time-honored traditions, we are invited to know ourselves and "seek within" for the answers to our deepest mysteries.

A new interpretation of these ancient invitations, and others, suggests that the key to solving the mystery of our origin, and discovering our life's purpose, is not waiting to be discovered in the world around us. *The answer is living inside of us.* It's safely locked away within the cells of our bodies, awaiting the day that we develop the awareness and the technology to read the message. Rather than sifting through the remains of crumbling temple walls and weathered manuscripts, the answer to our deepest mystery has been with us all along, preserved as the expression of life itself.

The moment that we allow ourselves to acknowledge the message in our cells, we are changed in one of two ways. We choose to either

- discount the message and ignore its meaning by writing it off as a fluke of math and linguistics; or
- shift our beliefs about who we believe we are, and the way we think of ourselves in the world to accommodate the existence of the message.

While the discovery of a message in our cells sounds amazingly hopeful, it also sounds like the stuff of science fiction. As much as we may want to believe that such a message exists, one of the first questions that typically comes to mind when we're presented with this possibility is "How?" How is it possible to write words that we can read, like the pages of a book, into the sticky, gooey stuff that forms the nucleus of a cell?

A peer-reviewed scientific paper in the *International Journal of Computer Applications* provides a surprising answer to this question.

## DISCOVERING HOW TO WRITE WORDS INTO DNA

In 2007, systems biologist and computer scientist Masaru Tomita, a professor at Tokyo's Keio University, published a paradigm-altering paper describing something that appeared to be straight out of a science fiction movie. The paper described how a group of scientists had written into, stored, and then later retrieved information from the DNA of a living organism.[1] The implications of the ability to write and read data that has been encoded into a living being are vast, and for some people, mind boggling.

These experiments, and others that have been performed since the discovery, show that complex information can, in fact, be written into and stored in the building blocks of a living being. They also demonstrate that the information can be preserved in the genome of that being for as long as it, or future generations that descend from it, continue to exist.

For a deeper understanding of precisely what I'm saying here, and to unpack the process step by step, I'll briefly describe the sequence of techniques that make this possible as follows.

**Step 1. Translating keyboard language to machine language.** When we type a message to a friend on our smartphone or from a computer keyboard, the familiar letters of our written language don't make sense to the computer's operating system. Early in the development of computer technology, a universal standard was adopted that would solve this problem for all keyboards, everywhere, for all time. The rules were put into place that made it possible to translate the alphabetic letters that we write into binary code that the computer can recognize for any language used today, or in the future. In 1963, that standard, the *American Standard Code for Information Interchange*, abbreviated simply as *ASCII*, was published.

Today, our computers, smartphones, and tablets all use ASCII to convert our typed messages into the unique series of ones and zeros, the binary code that represents each letter or number that we have typed. In order to store written words in a strand of DNA, it's this binary code that serves as the common denominator between our message and the chemical codes of life.

**Step 2. Translating binary code into genetic code.** As complex as living beings appear to be to the naked eye, the DNA for every form of life on Earth is made from different combinations of only four compounds, known as the *DNA bases*. The DNA bases—or *nucleotides*—are the foundation of all life and commonly represented by the first letters of their names. Cytosine is represented by a capital *C*, thymine by a capital *T*, adenine by a capital *A*, and guanine by a capital *G*.

We commonly see this shorthand representation of DNA in science fiction movies such as *Jurassic Park* (1993), where the genetic code of dinosaurs is revived and brought back to life after being preserved in the hardened sap (aka amber) of ancient trees for millions of years. DNA shorthand was even used as the title of a 1997 sci-fi thriller featuring Ethan Hawke, Uma Thurman, and Jude Law. The film *Gattaca* is a dystopian tale of using genetic code to optimize humans for desirable traits. And the outcome, as you can imagine, is not a good one, as it fails to allow for the power of human consciousness and epigenetics to transcend naturally inherited chemical patterns.

When it comes to storing information in DNA, once the binary codes of information have been converted into the chemical codes of life, they can then be inserted into the DNA of any organism to become a part of that being's genome. Only recently has modern science perfected the ability to successfully insert and delete sections of DNA with precision.

**Step 3. Inserting genetic code into DNA.** Following the conversion of binary code into the chemical code as described in the previous section, the scientists in Tokyo used a technology known as *gene editing* to actually insert the information into the DNA of the rod-shaped bacteria *Bacillus subtilis*. This particular strain of bacteria is commonly used in research due to its resilience to a range of environmental conditions, as well as its ability to replicate quickly, with a doubling time that can range from as little as 20 minutes to as long as 120 minutes. Multiple generations, and the information encoded into the DNA of those generations, can thus be documented in a relatively short period of time.

The technology that makes gene editing possible is colloquially known as *CRISPR*, which is shorthand for *clustered regularly interspaced short palindromic repeats*. CRISPR is the equivalent of a word processor for genetic material because it allows the user to see and edit DNA on a computer screen to optimize an outcome. CRISPR, which was first used on humans in 2016, is now legal in the United States, and it is being used to develop enhanced genetic features, such as a strong immune response to non-small-cell lung cancer.

**Step 4. Retrieving information stored in DNA.** After the Japanese scientists translated the words of the message into binary code, then converted the binary code into genetic code and inserted it into the living DNA, they had the ability to read the information they had stored. The team simply reversed the process of the original coding to retrieve the message from living bacteria. In other words, they took the existing chemical codes of the DNA, converted them back to the binary codes from which they originated, and then into familiar alphabetic letters that they were able to read on the computer screen.

The message that was first used to test the process was brief, yet it held historic information. It was Albert Einstein's famous and almost-universally recognized equation relating matter to energy, $E = MC^2$, which was originally published in 1905. The Japanese scientists retrieved the message intact after hours of the bacteria doubling at the rate of approximately 120 times every 60 minutes.

**PURE HUMAN TRUTH 15:** In 2007, the first message was written into, and later retrieved from, a living organism, proving that it's possible to write, store, and retrieve intelligent information in DNA.

Since Masaru Tomita and his team proved the concept of DNA data storage, research institutions throughout the world have jumped on the bandwagon to improve and expand the techniques and process. In 2012, for example, a team of scientists at Harvard Medical School reported in the prestigious journal *Nature* that they had successfully stored 5.27 megabits of information into a DNA molecule.[2] For perspective, this number represents over 600 times the amount of data previously stored in a living molecule.

## DNA: BETTER STORAGE THAN COMPUTER FLASH MEMORY

The discovery that DNA can be used as a storage medium is more than a curious oddity. A 2021 study reported by the National Institutes of Health (NIH) stated that the DNA molecule is actually better for information storage than the familiar flash storage that we typically use on our computers. And not just a little bit better. The paper stated that DNA is 1,000 times denser than the flash storage that is considered state of the art for our technology today.[3]

One reason for its density is because of the 3D nature of a DNA molecule. The information it holds is stored more efficiently than simply writing it out from left to right (or right to left in some languages).

Additionally, once the information is inserted into the DNA of a bacteria, or any form of life that is being used as the storage host, it doesn't require any additional energy to maintain the storage. Beyond the life energy used for respiration and metabolism, there is no additional energy needed to preserve the information.

Perhaps most importantly, when it comes to the topic of this book, the storage time does not appear to have a shelf life. The research paper states that once information is stored in the DNA of the bacteria, it becomes a part of its genetic blueprint—a living genome—and it will last as long as the organism carrying the genome exists.

It's precisely for this reason that I'm sharing this information with you.

**PURE HUMAN TRUTH 16:** DNA is more efficient as a storage medium than computer flash storage, making it a likely candidate for the location of an ancient message coded into our biology long ago.

Could the process of storing information in an organism's DNA be the human story as well? Is it possible that 200,000 years ago, when the first of our kind emerged in this world, our genome was modified to store information about us that would one day reveal the secret of our heritage? Our hidden potentials? And perhaps even our destiny?

If so, what does this mean for our lives today?

## IS THE SECRET OF OUR ORIGIN STORED IN OUR DNA?

The discovery of the DNA double helix was made only 70 years ago, in 1953. In the relatively short period of time that has followed, we've come light years in terms of our scientific understanding. If we've been able to achieve so much so quickly, what level of genetic information storage would be available to an advanced civilization with hundreds or thousands of years, or even tens of thousands of years, of technology under their belts?

In light of the discovery that information storage in DNA is possible, we owe it to ourselves to ask an obvious question: A long time ago, did a technologically advanced species intervene in our evolution and place us on the fast track to becoming the humans that we are today? And if so, did they leave a "signature" of some kind within us?

The answer to both of these questions is a resounding "Yes!" And it's now beyond theory that these things are a part of our evolutionary history. Just as works of art are signed by the artists who completed them, to proudly announce what they've accomplished, our bodies carry evidence of who (or what) was responsible for our existence.

Together, in the following sections, we'll derive the message that was encoded into our bodies at the time of our emergence. We'll also reveal the meaning and the power of the message that we've carried in each cell of our bodies since the first of our kind emerged 10,000 generations ago.

## THINKING DIFFERENTLY ABOUT CELLS

To read the message in our DNA, we need to think differently about cells themselves. In high school biology class, we were taught that the cell is the basic building block of human tissue. It's estimated that the average human body is made of approximately 50 trillion cells and that those cells are replaced at a rate of about 330 billion each day. These are mind-boggling numbers, and sometimes it's hard to wrap our minds around just what they represent.

When it comes to imagining one trillion cells, for example, it would take 32,000 years to count out a trillion seconds, one by one. Clearly, therefore, we have a lot of cells in our bodies, and each of those cells has a nucleus that contains our DNA as the 23 pairs of chromosomes that define our species.

Other forms of life have different numbers of chromosomes that define them, determined by the kind of life-form they are. Chimpanzees, for example, have 24 pairs of chromosomes, mice have 20 pairs, and fruit flies have only four pairs. Whether it's humans or flies, however, the basic templates of the chromosomes are the same. Each chromosome is made of long strands of DNA that are, in turn, made of shorter DNA segments called *genes*.

When we think of our cells from a biological point of view, we typically think of the sticky, gooey, gel-like substance that makes up the inside of the cell. It's difficult to envision how we could ever write and store information in this soft, wet stuff. To do so, we must shift our thinking a bit, and put on our information technology hats. From an IT perspective, the cells begin to take on a new look, offering us unconventional and exciting new possibilities.

For example, what if every cell of the human body functions as a library of information? And what if the chromosomes within the nucleus are actually "books" within the library?

If this analogy makes sense so far, then what comes next is expected. Within any book there is written material made of chapters, paragraphs, sentences, and words. And this is the function of the shorter genes: they serve as the chapters, sentences, and words of the genetic books.

Admittedly, this is a very different way of thinking of our biology.

**PURE HUMAN TRUTH 17:** Human cells may be thought of from an information technology perspective, with each cell being a library, the chromosomes being books, and the shorter strands of genes being chapters, paragraphs, sentences, and words.

While I use the library analogy to share the concept, the description is more than a metaphor. *It describes what is actually happening in our cells.* In a very real sense, the biology that gives us life is made of combinations of genetic codes that can be translated into meaningful words that we read just the way you are reading the words on this page right now. The question now is: "How do we read the message 'written' in our cells?"

The answer to the question is what follows, and is the result of 17 years of original research that required my willingness to look beyond the traditional boundaries that have separated ancient languages, modern science, biology, and linguistics in the past.

While I'll summarize this research here to reveal the message in our cells, the history and detailed derivation of this message is the subject of an entire 2004 book that I released through Hay House, *The God Code.*

*Note:* In the following six sections of this chapter, I'll share the details of the process that translates the chemical message stored in our DNA into words that we can read like the pages of a book. If you'd like to bypass the technical details of this process, you can go directly to Pure Human Truth 20 (see page 55), which reveals the message left for us by our creator. After that, we will begin exploring what this message written in our DNA means for our lives.

## FROM CHEMICAL DNA TO WRITTEN LETTERS

Previously, I described how the DNA of all life is made of four chemical bases, abbreviated as C, T, A, and G, which are technically known as DNA nucleotides. Each base is composed of combinations of the elements hydrogen, nitrogen, oxygen, and carbon. However, it's the number of elements within each DNA base—the count—that sets one apart from another.

Those differences are shown in the table below.

| Base Name | Abbreviation | Elements | Count |
|---|---|---|---|
| **Thymine** | T | Hydrogen<br>Nitrogen<br>Oxygen<br>Carbon | 6<br>2<br>2<br>5 |
| **Guanine** | G | Hydrogen<br>Nitrogen<br>Oxygen<br>Carbon | 5<br>5<br>1<br>2 |
| **Cytosine** | C | Hydrogen<br>Nitrogen<br>Oxygen<br>Carbon | 5<br>3<br>1<br>4 |
| **Adenine** | A | Hydrogen<br>Nitrogen<br>Oxygen<br>Carbon | 5<br>5<br>0<br>5 |

Hydrogen, nitrogen, oxygen, and carbon are all elements that can be found on the periodic table of elements. In the following rendering of this table, I've highlighted these elements so they may be located easily.

**Periodic Table of the Elements**

| 1 | 2 | 3 | 4 | 5 | 6 | 7 | 8 | 9 | 10 | 11 | 12 | 13 | 14 | 15 | 16 | 17 | 18 |
|---|---|---|---|---|---|---|---|---|---|---|---|---|---|---|---|---|---|
| 1 H HYDROGEN 1.008 | | | | | | | | | | | | | | | | | 2 He HELIUM 4.002 |
| 3 Li LITHIUM 7.0 | 4 Be BERYLLIUM 9.012 | | | | | | | | | | | 5 B BORON 10.81 | 6 C CARBON 12.011 | 7 N NITROGEN 14.007 | 8 O OXYGEN 15.999 | 9 F FLUORINE 18.998 | 10 Ne NEON 20.180 |
| 11 Na SODIUM 22.989 | 12 Mg MAGNESIUM 24.305 | | | | | | | | | | | 13 Al ALUMINUM 26.981 | 14 Si SILICON 28.085 | 15 P PHOSPHORUS 30.973 | 16 S SULFUR 32.07 | 17 Cl CHLORINE 35.45 | 18 Ar ARGON 39.9 |
| 19 K POTASSIUM 39.098 | 20 Ca CALCIUM 40.08 | 21 Sc SCANDIUM 44.955 | 22 Ti TITANIUM 47.87 | 23 V VANADIUM 50.941 | 24 Cr CHROMIUM 51.996 | 25 Mn MANGANESE 54.938 | 26 Fe IRON 55.84 | 27 Co COBALT 58.933 | 28 Ni NICKEL 58.693 | 29 Cu COPPER 63.55 | 30 Zn ZINC 65.4 | 31 Ga GALLIUM 69.72 | 32 Ge GERMANIUM 72.63 | 33 As ARSENIC 74.921 | 34 Se SELENIUM 78.97 | 35 Br BROMINE 79.90 | 36 Kr KRYPTON 83.80 |
| 37 Rb RUBIDIUM 85.468 | 38 Sr STRONTIUM 87.6 | 39 Y YTTRIUM 88.905 | 40 Zr ZIRCONIUM 91.22 | 41 Nb NIOBIUM 92.906 | 42 Mo MOLYBDENUM 96.0 | 43 Tc TECHNETIUM 97.907 | 44 Ru RUTHENIUM 101.1 | 45 Rh RHODIUM 102.905 | 46 Pd PALLADIUM 106.4 | 47 Ag SILVER 107.868 | 48 Cd CADMIUM 112.41 | 49 In INDIUM 114.82 | 50 Sn TIN 118.71 | 51 Sb ANTIMONY 121.76 | 52 Te TELLURIUM 127.6 | 53 I IODINE 126.904 | 54 Xe XENON 131.29 |
| 55 Cs CESIUM 132.905 | 56 Ba BARIUM 137.33 | 56-71 LANTHANOIDS | 72 Hf HAFNIUM 178.5 | 73 Ta TANTALUM 180.947 | 74 W TUNGSTEN 183.8 | 75 Re RHENIUM 186.21 | 76 Os OSMIUM 190.2 | 77 Ir IRIDIUM 192.22 | 78 Pt PLATINUM 195.08 | 79 Au GOLD 196.966 | 80 Hg MERCURY 200.59 | 81 Ti THALLIUM 204.383 | 82 Pb LEAD 207 | 83 Bi BISMUTH 208.980 | 84 Po POLONIUM 208.982 | 85 At ASTATINE 209.987 | 86 Rn RADON 222.017 |
| 87 Fr FRANCIUM 223.019 | 88 Ra RADIUM 226.025 | 89-103 ACTINOIDS | 104 Rf RUTHERFORDIUM 267.122 | 105 Db DUBNIUM 268.126 | 106 Sg SEABORGIUM 271.134 | 107 Bh BOHRIUM 274.144 | 108 Hs HASSIUM 277.152 | 109 Mt MEITNERIUM 278.156 | 110 Ds DARMSTADTIUM 281.165 | 111 Rg ROENTGENIUM 282.169 | 112 Cn COPERNICIUM 285.177 | 113 Nh NIHONIUM 286.183 | 114 Fl FLEROVIUM 289.191 | 115 Mc MOSCOVIUM 290.196 | 116 Lv LIVERMORIUM 293.205 | 117 Ts TENNESSINE 294.211 | 118 Og OGANESSON 294.214 |

| | | | | | | | | | | | | | | |
|---|---|---|---|---|---|---|---|---|---|---|---|---|---|---|
| 57 La LANTHANUM 138.905 | 58 Ce CERIUM 140.12 | 59 Pr PRASEODYMIUM 140.907 | 60 Nd NEODYMIUM 144.24 | 61 Pm PROMETHIUM 144.912 | 62 Sm SAMARIUM 150.4 | 63 Eu EUROPIUM 151.96 | 64 Gd GADOLINIUM 157.2 | 65 Tb TERBIUM 158.925 | 66 Dy DYSPROSIUM 162.50 | 67 Ho HOLMIUM 164.930 | 68 Er ERBIUM 167.26 | 69 Tm THULIUM 168.934 | 70 Yb YTTERBIUM 173.04 | 71 Lu LUTETIUM 174.967 |
| 89 Ac ACTINIUM 227.027 | 90 Th THORIUM 232.038 | 91 Pa PROTACTINIUM 231.035 | 92 U URANIUM 238.028 | 93 Np NEPTUNIUM 237.048 | 94 Pu PLUTONIUM 244.064 | 95 Am AMERICIUM 243.061 | 96 Cm CURIUM 247.070 | 97 Bk BERKELIUM 247.070 | 98 Cf CALIFORNIUM 251.079 | 99 Es EINSTEINIUM 252.083 | 100 Fm FERMIUM 257.095 | 101 Md MENDELEVIUM 258.098 | 102 No NOBELIUM 259.101 | 103 Lr LAWRENCIUM 262.110V |

The key to translating the chemical codes made from these elements into alphabetic letters is to find something they have in common: a link shared by both the letters and the elements. This is where we need to blur the boundaries between traditional sciences, such as chemistry, biology, and linguistics, to achieve a wisdom that is greater than the knowledge that an individual discipline can give us.

## WHEN LETTERS BECOME NUMBERS

The linguistic study of ancient languages reveals that the three alphabets that support the spiritual traditions of nearly half of the world's population, Sanskrit, Hebrew, and Arabic, are related. They each stem from the common hieroglyphic-like script *cuneiform*. Interestingly, cuneiform is not an actual alphabet made of discrete letters like we see in Hebrew, Arabic, and Sanskrit. Rather, it is a writing system that uses up to 1,000 characters to represent the syllables that make up a word. I'm mentioning this here because the message in our cells is so universal that it emerges even in the writing system that precedes the familiar alphabets of today's languages.

Saṇskrit is a language represented by the 49 letters of an ancient system known as Devanagari and remains in use today as the principal means of transmitting both traditional Hindu and Buddhist teachings. Arabic is classified as a Central Semitic language, and it is spoken and written in approximately 25 nations, as well as in Israel. The third language, Hebrew, is classified as a Northwestern Semitic language, and is commonly spoken in Israel today, as well as in traditionally Jewish households and synagogues throughout the world. An example of each writing system is shown below to give you a sense of what they look like and how different they are from one another.

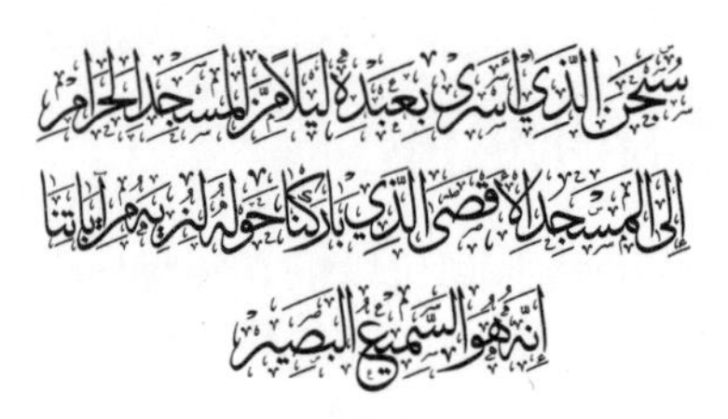


Examples of the four root writing systems, are cuneiform (top left), Sanskrit (top right), Arabic (bottom left), and Hebrew (bottom right). There's a lingering mystery shared by these four writing systems that has connected them in the past and continues to do so today. The mystery stems from the fact that each character of each script has a number that can be used interchangeably with the character to represent it.

**PURE HUMAN TRUTH 18:** The characters of cuneiform, Sanskrit, Arabic, and Hebrew have numeric equivalents that may be used interchangeably in written texts.

The study of these number/character/letter relationships is an ancient science, which, in the mystical traditions of the Kabbalah, has been formalized as the study of *gematria*. While each of the preceding writing systems are related and share the characteristics of gematria, only Hebrew, Arabic, and Sanskrit remain today as both spoken and written languages. The cuneiform that they stem from is no longer used and, for this reason, in the following sections I will focus upon the languages themselves rather than the scripts that they stem from.

As we see in the following Hebrew gematria table, the number for each letter is unique. It is constant and never changes. The source of the numbers, however, has long been a mystery.

| | | | | | | | | |
|---|---|---|---|---|---|---|---|---|
| א | Aleph | 1 | י | Yod | 10 | ק | Kof | 100 |
| ב | Bet | 2 | כ | Kaf | 20 | ר | Resh | 200 |
| ג | Gimel | 3 | ל | Lamed | 30 | ש | Shin | 300 |
| ד | Dalet | 4 | מ | Mem | 40 | ת | Tav | 400 |
| ה | He | 5 | נ | Nun | 50 | ך | Kaf (final) | 500 |
| ו | Vav | 6 | ס | Samekh | 60 | ם | Mem (final) | 600 |
| ז | Zayin | 7 | ע | Ayin | 70 | ן | Nun (final) | 700 |
| ח | Het | 8 | פ | Pe | 80 | ף | Pe (final) | 800 |
| ט | Tet | 9 | צ | Tsadi | 90 | ץ | Tsadi (final) | 900 |

To be clear, gematria is not the same as the practice of numerology that is common in some New Thought communities today. Rather, gematria is linked to the ancient science of Kabbalah, and is often referenced with regard to a 2nd-century manuscript attributed to the esteemed Rabbi Eliezer ben Hyrcanus of Judea.[4] Although the original text no longer exists, its contents are frequently mentioned in Jewish religious documents, and are known as the 32 rabbinical rules for interpreting the Bible.

The *American Heritage Dictionary of the English Language* defines science as "any methodological activity, discipline, or study."[5] In other words, for something to be validated as science, it must be consistently repeatable and predictable. With this definition in mind, the application of the 32 rabbinical rules to the study of gematria can be considered as a science, because it yields precise and repeatable outcomes from specific operations between letters, phrases, and words, and does so consistently.

## GEMATRIA REVEALS HIDDEN MEANINGS

Through the one-to-one relationship between the letters of the alphabets and their number equivalents, gematria reveals the hidden relationships, and deeper layers of meaning, that may not be apparent when reading only the words themselves. Regardless of any significance that has been assigned by language, culture, or society to particular words, the relationships revealed through numbers speak directly to the true, natural, and mystical meaning behind their letters. An example of these relationships may be seen in the gematria of the two words *heaven* and *soul.*

Almost universally, our most cherished beliefs suggest that the soul is the part of us that returns to the place of its origin after the death of the body and continues to live in the realm that we know as heaven. The traditional thinking is that this relationship cannot actually be proven, and is only implied through religious and spiritual writings and beliefs.

The direct link between *heaven* and *soul* revealed as a number, however, actually shows us this relationship, and it does so in a graphic format. The following analysis from the work of Rabbi Benjamin Blech, one of the great Hebrew scholars of our time, offers clarity, and for some people, a much-needed sense of comfort when it comes to the mysterious relationship between the realm of heaven and the human soul.[6]

Unlike English, where we are used to seeing vowels and consonants of the alphabet form our words, the traditional Hebrew language, when written, is composed only of consonants. While the vowels may be implied, and can be noted through special markings accompanying the letters, the biblical writing of Hebrew omits the vowels. With this in mind, the Hebrew word for *soul* is "NeShaMaH," which is written without the vowels as *N, Sh, M,* and *H.* Assigning the number codes of gematria described in the last section to these letters, we get the values of *N, Sh, M,* and *H* as follows:

| Hebrew Letter | N | Sh | M | H |
|---|---|---|---|---|
| Gematria | 50 | 300 | 40 | 5 |

The rules of gematria permit us to add these numbers to reveal a deeper meaning and allow us to compare this revealed meaning to other words that are also converted to gematria. When we add these individual values, as seen in the following example, the new, combined value becomes 395.

| Hebrew Letter | N | Sh | M | H | Sum |
|---|---|---|---|---|---|
| Gematria | 50 | 300 | 40 | 5 | 395 |

From the perspective of gematria, the number 395 is interchangeable with the word *soul.* This new, numeric representation of soul now makes it possible for us to explore a deeper relationship that is not immediately obvious in the written form of the word.

The word *heaven* is related to the 395 of the soul. To understand precisely how, we apply the identical process of assigning number values to the letters in it. In Hebrew, *heaven* is spelled as Ha-ShaMaYiM. As consonants only, the word is written as follows:

| Hebrew Letter | H | Sh | M | Y | M |
|---|---|---|---|---|---|
| Gematria | 5 | 300 | 40 | 10 | 40 |

Representing each letter as its numeric value, we discover a direct and perhaps surprising relationship between the two words *heaven* and *soul*. Adding the number values for *heaven* yields the number 395—*precisely the same value* as the word *soul!* Their numeric values are identical.

| Hebrew Letter | H | Sh | M | Y | M | Sum |
|---|---|---|---|---|---|---|
| Gematria | 5 | 300 | 40 | 10 | 40 | 395 |

One of the rules for the science of gematria is that two words related in number are understood to be related in meaning as well. This implies that, because they are one and the same, the big realm of heaven embodies something that is alive within us, and we, in turn, are more than our physical bodies and are intimately enmeshed with the heavenly realm as well.

In the traditional biblical descriptions of our origin, "God created the heaven and earth" as two distinct yet related realms of experience. Other mystical texts, such as the ancient Haggadah, which contains Kabbalistic writings, as well as the early Talmud, describe humankind as a bridge between soul and heaven, stating, "He [humankind] unites both heavenly and earthly qualities within himself . . ."[7]

It's through our existence that the qualities of the two realms are merged into a single expression. At the end of our lives, the heaven and earth that have merged as our bodies return to their respective abodes. Our bodies return to being the dust of the earth, while our souls and heaven are already one. This perspective of life and death offers a tangible sense of why the experience of death does not mean an end to our existence. The hidden number code assures us that it is so.

## NATURE'S MYSTERIOUS NUMBERS

The rules of gematria also permit us to add the numbers representing letters in order to reduce them to simpler values between one and nine. It's this reduction process that also reveals even deeper meaning, as the simpler numbers are closer to nature and nature's systems. There are a number of terms for this process that stem from the different traditions, including Pythagorean addition, Kabbalistic reduction, and in more modern vernacular, decimal parity.

Present-day mystical traditions, including those of the Freemasons, apply this form of spiritual mathematics to reveal deeper meanings for life, family, and world events. Former Masonic Lodge Master Peter Taylor describes the significance of the numbers one through nine as he relates them to the Greek and Egyptian traditions, as well as the Christian angelic hierarchy. "These 9 gods/angels were archetypal principles that regulated and ruled the cosmos through the laws of number."[8]

In the case of the number 395, representing the words *heaven* and *soul*, the reduced value becomes the simpler number eight, falling between the natural numbers of one and nine and the archetypal principles they represent. (The math would be $3 + 9 + 5 = 17$ and $1 + 7 = 8$.) The reduced number still shows that both words have the same value, and therefore are equivalent to one another in meaning.

While it is not necessary to reduce the values of word numbers over 10, it is done under some circumstances to explore deeper relationships between words. This will be apparent in the following sections when we apply the rules of gematria to the periodic table of elements and the message embedded in our cells.

With the ideas of gematria and the 32 rabbinical rules in mind, when we apply these principles to the message coded into our DNA, we discover that it's possible to convert the elements of the DNA into the familiar letters of the related ancient alphabets. For clarity, it is important to note that the DNA elements translate into exactly the same words, with exactly the same meaning, in each of the related alphabets, as well as in the root script from which they all stem. The message literally reads exactly the same in cuneiform, Sanskrit, Arabic, and Hebrew.

When we think about this, it should come as no surprise. It would make no sense to trust the powerful and unifying message of humankind's origin to a single language, a single religious text, or people of a single ethnicity. It stands to reason that the message would be found in all four of the ancient root languages.

Because the Hebrew language is the basis of the Abrahamic texts, and those texts are the foundation of Islam, Christianity, and Judaism—the religions practiced by two-thirds of the world's population—for the purposes of this book we'll translate the ancient DNA code from the perspective of the Hebrew alphabet.

## DISCOVERING THE LINK BETWEEN LETTERS AND NUMBERS

In a previous section, we discovered that the four DNA nucleotide bases are made of hydrogen, nitrogen, oxygen, and carbon, each of which appears on the periodic table of elements (see page 45). Although these four elements are each unique in their properties, there are many different ways that we can represent them using both words and numbers. When we identify them using words, for example, we use their familiar names in whatever language we happen to speak. In English, those are hydrogen, nitrogen, oxygen, and carbon.

When we represent them using numbers, however, the process is not so cut and dried. There are many different numbers representing different properties, such as atomic number, atomic mass, melting point, valence, and so on, that we could assign to represent each element.

During the time that I was researching this material, personal computers were less common, the Internet was in its infancy, and it was a tedious manual operation to explore each of the properties linked to the elements in search of a single number that could represent the gematria of the letters of the ancient alphabets. The research revealed that there exists one, and only one, number that links the elements of our DNA with the gematria of words spelled out by ancient alphabets. That number represents the property known as *atomic mass.*

*Mass* is an interesting concept that represents how much space something occupies rather than how much it physically weighs. In hindsight, it makes perfect sense that atomic mass would be the common thread between the elements and the DNA, as the concept of mass is precisely what the message in our bodies is telling us. It identifies how we exist in the space we occupy while living in a physical world.

Now that we know that atomic mass is the key to decoding the message in our cells, we can build a chart to correlate the elements of the DNA bases with the numbers representing the letters of the Hebrew alphabet.

**PURE HUMAN TRUTH 19:** The unique numbers that represent the atomic mass of each of the four elements that compose our DNA bases can be added up to total numbers that form various words written in the ancient root languages.

In the table below, we see the simple layout of the numbers linking the letters of the Hebrew alphabet with the four elements. *Note:* These correlations would work equally well for all four of the root alphabets: cuneiform, Sanskrit, and Arabic as much as Hebrew.

| DNA Element | Reduced Atomic Mass | Reduced Gematria | Matching Letter from the Hebrew alphabet |
|---|---|---|---|
| H | 1 | 1 | Y |
| N | 5 | 5 | H |
| O | 6 | 6 | V |
| C | 3 | 3 | G |

With this chart, we can now replace the elements of DNA with the letters of the alphabets. And when we do, something amazing, beautiful, and sacred unfolds before our eyes.

## THE MESSAGE IN OUR DNA TRANSLATED

What happens if we replace an element of the DNA bases, C, T, A, or G, with the equivalent letter of the Hebrew alphabet? The tables below show the gematria of the DNA base cytosine before and then after translation.

### The DNA Base Cytosine before Translation

| **DNA Base Name:** | **Cytosine** | | | |
|---|---|---|---|---|
| Elements: | H | N | O | C |
| Count: | 5 | 3 | 1 | 4 |
| | H | N | O | C |
| | H | N | | C |
| | H | N | | C |
| | H | | | C |
| | H | | | |

### The DNA Base Cytosine after Translation

| **DNA Base Name:** | **Cytosine** | | | |
|---|---|---|---|---|
| Elements: | H | N | O | C |
| Count: | 5 | 3 | 1 | 4 |
| | **Y** | **H** | **V** | **G** |
| | Y | H | | G |
| | Y | H | | G |
| | Y | | | G |
| | Y | | | |

Immediately, we begin to see the power of the message revealed in this translation. The information coded into our DNA appears to have been done so in layers, just as the chapters of a book unfold in a sequence. The message that we're decoding in this process is the equivalent of reading the Introduction (the first layer of code) in the DNA book of our cells. In the translation chart, we see that the highlighted letters of the message read *YH VG*. These are two words, represented as consonants only, that read as follows:

*YH translates to: "God/eternal"*

*VG translates to: "Within the body"*

With a clarity that is rare when it comes to the discussion of human beginnings, the translated message leaves no doubt as to the source of our origin. Beyond metaphor or speculation, the DNA that makes our bodies and our lives possible is literally telling us that we humans are God/eternal within the body.

The message is concise. It doesn't tell us who or what God is. It doesn't tell us where God comes from. Just as Vincent van Gogh's signature on his famous painting *Starry Night* tells us nothing about where he lived or what his beliefs were when he created this masterpiece, the "signature" embedded in the strands of our DNA simply exists as an acknowledgment from the architect of our genome: *YH*, the personal name of God as it is written in the oldest records of the Dead Sea Scrolls, the Old Testament, and the Torah, and "within the body" revealing where the architect's work resides.

While our spiritual traditions ask us to solve the mystery of our beginning through the stories passed from father to son, mother to daughter, generation to generation, the direct evidence supporting the stories has required an act of faith. That is, until now.

**PURE HUMAN TRUTH 20:** When we perform the substitution of letters for their numeric equivalents from the atomic mass of our DNA, the first layer of the code for each cell of our bodies reads as: "God/eternal within the body."

When we apply the identical process of converting the atomic mass of the elements making our DNA to the letters of the ancient alphabets for the remaining genetic bases, we reveal the same message, expressed in varying degrees, encoded in each of the DNA bases for life. The next table illustrates the results of doing so.

| Layer 1 DNA Bases Translated | | | | |
|---|---|---|---|---|
| **DNA Base Name** | Cytosine | Guanine | Thymine | Adenine |
| Element Name | HNOC | HNOC | HNOC | HNOC |
| Element Count | 5314 | 5515 | 6225 | 5505 |
| | **YHVG** | **YHVG** | **YHVG** | **YH G** |
| | **YH G** | **YH G** | **YHVG** | **YH G** |
| | **YH G** | **YH G** | Y G | **YH G** |
| | Y G | **YH G** | Y G | **YH G** |
| | Y G | **YH G** | Y G | **YH G** |

The translation of the elements that make up the code of life in the four DNA bases into the letters of the Hebrew alphabet reveals that our bodies are literally made of different combinations of the ancient name of God, YH. Cytosine, for example, reads as "God/eternal within the body; God/eternal; God/eternal."

Guanine reads as "God/eternal within the body; God/eternal; God/eternal; God/eternal; God/eternal." And so on for the remaining translations of the first letter in the English name for the elements in the nucleotide bases thymine and adenine.

## IS IT COINCIDENCE?

One of the first questions that arises when we reveal the results of such a process is with regard to the chances that, while unusual, it is possible that the results are just lucky maneuvering and a trick of numbers. What are the odds that the entire message is a fluke?

This is a good question, and I asked myself the same question early in the process, the first time the message jumped off the page of my legal pad of paper as I was working with my late-night, handwritten calculations.

A statistician friend of mine answered the question quickly. The odds of randomly forming the four letters of two meaningful words (YH VG) from the 22 letters of the Hebrew alphabet are .00042 percent, or 1 in 234,256. While these are certainly not astronomical odds, like 1 in a bajillion, they tell us that the probability of the message we carry in our DNA is beyond chance. In other words, the message is intentional.

**PURE HUMAN TRUTH 21:** The statistical odds of the message "God/eternal within the body" forming within our DNA by chance are .00042 percent, telling us that the probability of us carrying a message of such significance is beyond chance.

And when we factor in the meaning of the two words that are formed—God/eternal within the body—we see immediately that there is an intentionality underlying the presence of the message itself. It's precisely because there is an intention, as well as a meaning, that a purpose is implied. What could be the purpose of leaving a message in the building blocks of our lives, and doing so in a way that we could only read it when our technology had advanced to the point of allowing us to do so? Maybe this is precisely the point.

## A MESSAGE FOR US TODAY

Maybe whoever or whatever is responsible for our existence coded the message into our cells for precisely this moment in our history. It's only now, when we have advanced computers, that we can map the human genome. It's only now that we can code information into, and retrieve information from, the cells of a living organism. It's only now, when we have CRISPR technology, that we can edit DNA and change the biology of an embryo while it is still living in the womb of its mother. And it's only now, when the world is perched on the precipice of war and widespread destruction, that we have cracked the code of life and unleashed the power of the atom in ways that can be used for good or evil applications.

It's more than a simple coincidence that only now, with the convergence of these vital factors, and others, that we are able to discover a message so universal that it transcends any differences we could ever use to divide us.

Regardless of the color of our skin, regardless of our religion, beyond the differences that we may have in sexual preference, the pronouns we prefer to describe ourselves, and the gender we identify with, the message reads exactly the same for every child, woman, and man that walks the Earth today, or that ever will in our future.

When we find ourselves inundated with the biases of media and politics attempting to convince us that we are hopelessly flawed, we need look no further than the 50 trillion cells in each of our own bodies to be reminded of who we are. We are God/eternal within the body.

The transhumanist technology that is being imposed upon us by powerful factions of our society threatens to alter our genetic essence, and in doing so, to forever alter the sacred message left with us long ago.

## THE PURE HUMAN BODY IS A TEMPLE

We can only carry a full expression of the message in our DNA when we are pure humans, absent of synthetic materials, chemicals, and genetic edits that alter our natural genome. When we give ourselves away to the technology, in addition to eliminating the biological antenna that links us to our divinity, we are also altering and deleting the ancient message of our identity that was left within us long ago.

Some of the most cherished scriptures from world religions remind us that our bodies are "temples" and that it is within our physiological temples that we carry something sacred and precious in this world. When we allow ourselves to view the Christian Bible as a historic document, setting religion aside, for example, in the passage from First Corinthians, Chapter 3, Verse 16, the Apostle Paul reminds us: "Know ye not that ye are the *temple* of God, and that the Spirit of God dwelleth in you?" (author's italics)[9] In Chapter 6, Verse 19, of the same book, Paul removes any doubt about the meaning of his statement, declaring: "Know ye not that your body is the *temple* of the Holy Ghost which is in you?" (author's italics)[10]

While the idea of the body as a temple is commonly used as a metaphor by many people, the revelation of a direct message coded into our DNA gives new meaning to the idea of our bodies as sacred vessels. We are, in fact, biological temples. Within the holy of holies that is the nucleus of each cell, we carry the sacred DNA library of knowledge and the secret of our identity.

In 1991, scientists published the papers acknowledging that the fusion of our DNA and subsequent mutations, such as the occurrence of chromosome 2, was accomplished in a way that cannot happen in nature. They say the fusion that gives us our humanness is impossible under natural conditions. Chromosome 2 is only one example of such mutations that give us our human qualities.

These mutations are the product of an intentional act, and this is the discovery that makes the messages in our cells so powerful. We are walking vessels of advanced biology and engineering, poised at the precipice of giving away the very gifts that we cherish in our species. Perhaps our highest form of mastery is to honor and protect the vessel that carries the message of our identity—to honor the gift of our pure humanness.

## HOW DO YOU FEEL ABOUT THE MESSAGE THAT LIVES INSIDE YOU?

Once you know of the message coded into the DNA of your existence, you can't unknow it. If you've read the words of the message printed in the pages of this book, those words are now etched into your awareness. If you've heard the message while listening to the audio version of this book, the sound of the words is also now a part of your knowing.

Once you are aware of the message, it's impossible to press the rewind button and go back to the moment that existed before you read or heard it. And once you know the secret of your past, the origin of our species, and the potential you carry within you, one of two things has to happen. You will either:

- Discount the message altogether and pretend that you never discovered it.
- Rearrange your beliefs and lifestyle to make room for what the message means to you.

Let's look at these two possibilities in turn.

## DISCOUNTING THE MESSAGE

You may choose not to believe what you've been shown. It may sound just too outrageous. You may attempt to invalidate what you've read by asking an AI system such as Gemini or ChatGPT to confirm the message. (They will not be able to do so, because the AI draws upon existing information, as it is filtered through community guidelines and media filters that acknowledge only commonly accepted information, for its replies.) You may also ask your friends and family if they've ever heard of such a thing. The implications of having God's ancient name coded into every cell of your body may feel so overwhelming that it's just too much to truly embrace.

Just the way we typically discount bad news when it comes to us unexpectedly, you may simply turn to the next page of the book, or close the book altogether and place it on a shelf with your collection of other self-help books that have also attempted to convey to you the deep truth of your inner power. And to do any of these things is a perfectly normal and perfectly natural thing to do. In psychological terms, it falls under the heading of *normalcy bias*.

Normalcy bias is what happens when we attempt to minimize the impact of something that threatens the belief system that we've created to make sense of the world, and our lives. It's a healthy defense mechanism that smooths the rough edges from disturbing and disruptive information or a threatening situation.

The comfort of the normalcy bias is often temporary, however, because it doesn't change the reality of what's happening in our lives; it only changes the way we feel about what's happening in our lives. When something is true, it will appear in many places, in many ways.

At some point, the fact that an intentional message is coded into the essence of your existence will show up in your life. And when it does, you'll need to make the choice to embrace, or to deny, the implications of the message.

## EMBRACING THE MESSAGE

For many people, the message "God/eternal within the body" is a welcome and joyous communication that only confirms what they've always suspected. Seeing the process of the message derived through number and letter correlations simply helps their left brain accept what their right brain, and their heart, has always known to be true.

Once the message is accepted, the question becomes, "What does it mean in your life?" How differently will you live your life, solve your problems, respond to challenges, and heal your body after learning that you are "God/eternal within the body"?

When I ask our audiences this question during live seminars, while the answers vary from person to person, they all fall under the same umbrella. Almost universally, participants mention three things that they feel.

- A sense of empowerment and possibility. One of the first things that people tell me is that they now have a reason to think differently about themselves. They have something tangible to hang their sense of unrealized potential upon.

- A sense of sacredness when it comes to their own bodies. Knowing that a greater intelligence—regardless of who or what they believe that intelligence to be—left a coded message that lives as the very fabric of the body that gives them life, provides a reason to care for themselves in ways they may have neglected to in the past.
- A desire to defend the integrity of their bodies. It gives them a reason to consider carefully how they treat their bodies: what substances they introduce *into* their bodies and what they do and place *upon* their bodies. It gives them the reason to question if the alterations or the substances they are considering honor the message in their cells, and the gift of their lives.

All three of these feelings are messages. And sadly, these messages are lacking in our public education systems and healthcare policies. Knowing that we're God/eternal within the body tells us that, while we're definitely "in" this world, we're not "of" this world.

**PURE HUMAN TRUTH 22:** When we become discouraged, distracted, or deceived, or we simply have forgotten who we are, we need look no further than the 50 trillion cells within our bodies to be reminded that we are literally God/eternal within the body.

Perhaps the greatest level of mastery that we may attain for ourselves is discovering how we may best care for, protect, and preserve the vessel that holds the ancient message of our remarkable origin.

CHAPTER THREE

# Transhumanism

## Erasing the Link between Human and Divine

Technology is a useful servant, but a dangerous master.

— Christian Louis Lange (1869–1938),
Norwegian historian and political scientist

In my school years between the 1950s and the early 1970s, I was taught that the human body is a flawed form of life. High school science teachers taught that, from the moment of our first breath, we are a weak, vulnerable, and mostly powerless form of life when it comes to our relationship with the world around us, as well as the world within us.

The thinking was that, due to our flaws, we're incapable of meeting the challenges of life using our own natural abilities in a healthy way. And the conclusion drawn from this line of thought was that we need a savior. We need something outside of us to level the playing field.

Perhaps not coincidentally, it was during the time that these ideas were popular that the world was also in the midst of a technological revolution that reinforced the victim mindset. Powerful advances in science and technology had just made possible the landing of the first humans on the moon, the discovery of the double-helix structure of DNA, and the development of miniaturized circuits that reduced the size of computers that once filled an entire room to a size that fit neatly onto a desktop.

The magnitude of these advances led to an expectation that the pace of scientific discoveries was unstoppable. Perhaps not surprisingly, it was also believed that technology would solve the problems of the world, and our lives, forever. That technology would be our savior!

## NEW TIME, SAME THINKING

Decades have passed since my school-age experiences. And although we live in a different time, when it comes to the role of technology in our lives, the thinking hasn't really changed so much. In public schools today, we still teach our children that they're the product of random mutations and lucky biology, a theory that continues to be perpetuated even though it is no longer supported by the evidence. Additionally, our children are led to believe that their mere human capabilities are no match for microprocessors and advanced AI, and that they will always be powerless victims of a cruel and scary world.

It's because of these seemingly unfair circumstances that our young people are also being led to believe that their future and their "salvation" lie with the computers and an artificial intelligence that can level the landscape of life's disparities and "fix" the flaws of their human vulnerabilities. This is the kind of thinking that has brought us to the dangerous crossroads where we find ourselves today. Technological developments in miniaturized nanocircuits, AI, gene editing, and mRNA vaccine platforms have converged into a sweeping social movement encouraging the widespread replacement of our natural biology with synthetic organs, tissues, and artificial substitutes for our finely tuned bodily systems.

While the general idea of using human-made substitutes and gadgets to upgrade our bodies is nothing new, especially when it comes to the matter of replacing missing limbs or damaged organs, it's the sudden advances in synthetic intelligence, nanotechnology, and human-digital interfaces that have allowed this thinking to be expressed in ways that were unthinkable only a few short years ago.

Who would have thought, for example, that the U.S. Food and Drug Administration (FDA) would approve computer chips to be implanted into our brains so that we can have wireless communication with our iPads, smartphones, and computer hard drives? Or that a gene-engineered segment of DNA called *messenger RNA* (mRNA) would be introduced—even mandated to be introduced—into millions of human bodies worldwide to program their immune systems to focus upon preventing infection by a specific virus? Yet these events, and more, have happened recently. And they are not isolated instances.

There is an overt marketing campaign to normalize the acceptance of computer and radio frequency identification (RFID) chips into the body specifically, and the merging of life-altering technology into the human body in general. This campaign is part of a broader movement commonly known as the transhuman movement, or simply *transhumanism*.

As we'll discover in the following sections, the danger of taking a transhumanist approach to our bodies is that it places us squarely on a path of false evolution, leading to the loss of our humanness and our most cherished human traits and qualities.

## TRANSHUMANISM

The Latin prefix *trans-* means "beyond." It follows, then, that the term *transhuman* implies something that is beyond a natural human: a post-human form of life. Transhumanism, as a philosophy, views carbon-based life in general, and human life specifically, as defective by its very nature.

Among the attributes that are considered "flaws" are our advanced capacity for emotion, which can sometimes be "messy" and cloud our decision-making; our need for physical intimacy—sex—to bring new life into the world, along with the imprecise and unknown outcomes that follow birth; the belief that our intelligence is inferior to programmed and artificial intelligence; our vulnerability to disease and what's often thought of as the universal disease of old age, and ultimately, of death itself. Typical transhumanistic thinking concludes that, because these perceived vulnerabilities exist, we need technology to "fix" our flaws and save us from our suffering.

**PURE HUMAN TRUTH 23:** Transhumanism is a philosophy that advocates incorporating AI, computer chips, and electronic sensors into the human body to "fix" the flaws of our natural biological functions.

Irish author and journalist Mark O'Connell summarizes this reasoning: "It is [*the transhumanists'*] belief that we can and should eradicate ageing as a cause of death; that we can and should use technology to augment our bodies and our minds; that we can and should merge with machines, remaking ourselves, finally, in the image of our own higher ideals."[1] (emphasis is mine)

The rapid advance of extraordinary technologies never available to us in the past, such as gene editing and mRNA platforms, is pushing both the narrative that we need technology and the policies that allow these technologies to be implemented in our bodies on a mass scale.

## NOT YOUR PARENT'S TRANSHUMANISM

Transhumanistic thinking is nothing new.

For centuries, humans have considered ways to mimic the capabilities that we see in other forms of life so we can improve and enhance our own natural abilities. Roman warriors, for example, copied the protective, thick skin of wild elephants by covering themselves with thick leather and overlapping strips of pounded iron to protect them in battle. The 9th-century engineer Abbas ibn Firnas built artificial wings of lightweight bamboo and silk cloth to imitate the wings that make flight possible for wild birds. And while he did successfully fly with his machines for up to 10 minutes at a time, an unfortunate miscalculation in the way he built the vehicle's landing gear resulted in the serious injuries that put an end to his days of flying.

In more recent times, the use of wooden pegs to replace lost limbs and wooden teeth to replace those lost to disease or in combat are examples of attempts to artificially enhance the human body. I mention this to illustrate that the basic philosophy of transhumanism is not new. What is new, however, is the technology that allows us to carry transhumanistic thinking to never-before-seen levels that actually change who we are, rather than enhance what we can do.

Early in the 20th century, a number of academics, scientists, and philosophers laid a foundation of principles that led to the formalization of transhumanist philosophy. In an essay published in 1923 titled "Daedalus: Science and the Future," the early geneticist J. B. S. Haldane recognized that the emerging age of technology would catapult humankind beyond the limitations of our natural abilities. He also recognized that attempts to alter our biology would likely be met with resistance due to what many people would perceive as "indecent and unnatural" innovations.[2] I'll use the following discovery of the ability to customize the genes of living human embryos as a perfect example of what Haldane meant here.

In 2018, the scientific journal *Nature* announced the use of gene-editing technology in a way that caught the scientific community off guard and took the societies of the world by surprise. While the ability to edit genes for research purposes has been used since 1985, and the advanced gene editor known as CRISPR was introduced in 2006, until the time of the report in *Nature*, gene editing was limited to nonhuman research. It had occurred only with animals in research laboratories.

The article in the scientific journal put the world on notice that there has been a paradigm shift when it comes to the use of gene editing to customize a human life. The title of the article tells the whole story: "Genome-Edited Baby Claim Provokes International Outcry."[3] The title also highlights the controversy surrounding transhumanist philosophy: Do we humans have the right to modify and "improve" nature and the natural process of human life—yes or no?

## EDITING A HUMAN LIFE

The report in *Nature* described the birth of the world's first genetically edited human babies: twin girls who were the result of gene modification performed on them while they were still embryos. The scientist who made the announcement, He Jiankui, Ph.D., is a biophysicist from China's Southern University of Science and Technology. He stated that the purpose of the gene editing was to introduce a new feature into the genetic code of the girls, one that had never before been seen in humans: a resistance to infection from the HIV virus.

While the gene editing appeared to be well intentioned, and by all reports was successful, the global scientific community immediately came down hard upon Jiankui, chastising him for what he'd done. The criticism revolved around the as-yet-unresolved question of whether or not we have the moral right to customize a human fetus and influence the life that will follow—something that in the past has been left to nature, natural processes, and God.

**PURE HUMAN TRUTH 24:** Although the first successful gene editing of two human embryos in 2018 was illegal, it demonstrated that the process is beyond theory. It is now possible to successfully engineer human DNA in the womb after conception.

Commenting on the report, the head of the Scripps Research Translational Institute in the United States, cardiologist Eric Topol, M.D., said, "This is far too premature. We're dealing with the operating instructions of a human being. It's a big deal." [4]

In even harsher terms, cardiologist Kiran Musunuru, M.D., a gene-editing expert from the University of Pennsylvania, stated that what Jiankui had done was "unconscionable . . . an experiment on human beings that is not morally or ethically defensible."[5]

I'm sharing the story of the gene-edited twins to illustrate the point made by the scientific community. It goes to the very core of a larger debate that's been raging since the birth of science, and the scientific method, nearly 300 years ago. It all comes down to the practical applications of scientific discoveries, and to when we should, and when we should not, apply these discoveries to human life in the real world. The case of the Chinese scientist gives a face and a name to this debate. It also brings the conversation from the world of philosophy and academics into the real-world lives of everyday people.

The report in the scientific journal was a game changer. It forced what had previously been an interesting academic discussion of "what ifs" to go beyond academia and into the stark reality of formulating a real-world response to something that had actually occurred.

Rather than a discussion based on PowerPoint slides at a conference, there were now human lives at stake—two baby girls—whose genetic code had been manipulated before they were born in a way that would change their lives forever. The dilemmas that the scientific community, and now the world at large, face are simply: Do we have the right to do what the Chinese scientist did, and even more? Do we have the right to customize nature and the natural code of a human life to produce a desired outcome?

## BECAUSE WE CAN, DOES IT MEAN WE SHOULD?

Between 1976 and 1991, I was privileged to work as a problem solver in Fortune 500 companies alongside some of the leading scientists of the day. The expertise of those that I was surrounded by on a daily basis included the fields of earth sciences, physics, computer sciences, and space engineering. To address specific problems of the day, we were divided into teams that had access to the most advanced technologies, being applied to the most futuristic applications the world had ever seen.

From the Star Wars Defense Initiative (SDI) and the newly developed U.S. Space Command to advanced communication systems, space-based laser beams, and remote-sensing systems using the first desktop microcomputers, it was as if the proverbial floodgates had opened for innovation, new technology, and the funding of future science. Not surprisingly, along with the new discoveries came the discussions of how, and if, the gadgets that were being developed should ever be used.

This was especially true of the advanced weapons systems that were being developed as part of the Cold War that was raging at the time between the United States and the former United Soviet Socialist Republic (USSR), now Russia. The degree of responsibility that comes with developing technologies with the potential to affect human life, and to do so in a big way, sparked heated debates about the morality of implementing the technologies in the real world, and our right as a society to do so—debates that I enthusiastically joined at every opportunity.

The discussions were passionate, often heated, and generally followed one of two lines of thought. The first path emphatically stated that the fact that the discoveries had been made was, by default, the license to use what had been discovered. The logic was that if we were not meant to have and apply the technology in our lives, we would never have had the insight and good fortune to discover it. This way of thinking was often summarized using the adage "because we can, we should."

The second path of discussion took a very different view. While it also fully supported the exploration and discovery of new technologies, the second path differed in the thinking regarding how the innovations should be applied. This perspective often began with the words, "Hold on! Not so fast. Just because we *can* do something doesn't necessarily mean that we *should* do that something."

To the dedicated scientists and engineers who were advocates of this second way of thinking, the forces of nature represent sacred laws that should not be tampered with. To customize the genetic code of our children before they're born, for example, or to "adjust" global weather patterns to suit our agricultural or military needs; to use the universal energy packed into the quantum vacuum of space as a weapon that could destroy entire populations is off-limits, they argued. To do these things would violate an ancient and unspoken trust between us humans and any higher power that we subscribe to. We must first consider the implications and the consequences of the new technology before we use it and make it available to the world.

A historic example of this debate was revealed when the final documents for the Manhattan Project were declassified in 2014. The documents revealed that the scientists, while still working in isolated labs on compartmentalized aspects of the supersecret project, eventually began to realize that they were building a weapon—an atomic bomb—unlike anything the world had seen. Many insisted that while the weapon they were building could be used to demonstrate a principle, it should never actually be used on people. One of the most prominent of those scientists was the inventor of both the nuclear reactor and the electron microscope, physicist Leo Szilard, Ph.D. On July 17, 1945, Szilard wrote a letter to president Harry S. Truman on behalf of himself and many of his colleagues in which he stated the following:

> *Atomic power will provide the nations with a new means of destruction. The atomic bombs at our disposal represent only the first step in this direction and there is almost no limit to the destructive power which will become available in the course of this development. Thus, a nation which sets the precedent of using these newly liberated forces of nature for purposes of destruction may have to bear the responsibility of opening the door to an era of devastation on an unimaginable scale.*
>
> *In view of the foregoing, we, the undersigned, respectfully petition that you exercise your power as Commander-in-Chief to rule that the United States shall not, in the present phase of the war, resort to the use of atomic bombs.*[6]

Unfortunately, Leo Szilard's petition, signed by 70 scientists working on the Manhattan Project, went unheeded by the president. As history shows, on August 6, 1945, the weapon developed by the scientists at the Los Alamos National Laboratory in New Mexico—the world's first atomic bomb—was used against a civilian population in the Japanese city of Hiroshima with horrific and devastating consequences. The conversations I was part of were taking place at the height of the Cold War in the mid-1980s, only 40 years after atomic bombs were dropped on Hiroshima and, three days later, on Nagasaki.

During the post–World War II years, the combined nuclear arsenals of the world had grown to a total of over 60,000 atomic weapons. If they were ever used, they held the potential energy to microwave the entire planet many times over. It was for this reason that the threat of a nuclear war loomed very real at the time, and the conversations regarding the topic were typically heated and passionate.

While we're now in the post–Cold War era of the 21st century, and the Cold War is technically over, the argument that I'm describing regarding how technology is used is far from over. The difference between then and now is that today, rather than new discoveries being applied to weapons that destroy the world, the new discoveries are being applied to technologies that change or replace the human body—they're being directed at us—in new and unprecedented ways. The controversy regarding the morality of using AI, computer chip implants in the brain, and the gene editing of the human body are just the beginning of this debate.

## THE FUZZY LINE BETWEEN HUMANS AND MACHINES

Some scientists instinctively draw upon Aldous Huxley's book *Brave New World* to articulate their concerns when it comes to allowing technology to dominate our lives. Based upon the current trends and human nature, the frightening scenarios that are fiction in the book could easily become the reality of our lives. For example, in a single sentence neuroscientist and author Abhijit Naskar concisely sums up the danger of artificial intelligence in our lives, stating: "AI research can have irreversible repercussions in the life of the human species, so we must tread cautiously."[7]

The caution that Naskar advocates for is precisely the crossroads that this book is dedicated to. In our time of exponential growth in technology, and the choices we make regarding how we apply the innovations in our lives, we're in uncharted territory. In many respects, the transhumanistic innovations are happening faster than the thinking regarding the technology, what it means to us, and the morality of how it is applied in our lives.

Clearly, it's the thinking that justifies replacing natural biology with artificial components that distinguishes between a loving enhancement to restore war-damaged individuals to wholeness and a wholesale practice that is adopted as a mandatory, high-tech "upgrade" to all humans, in all nations, throughout the world.

There are some instances, as we'll see in the following section, where using artificial components, such as brain/chip prosthetics, can give someone who has lost limbs on the battlefield of war or in an industrial accident the gift of freedom in their lives and the ability to hold their babies in their arms, brush their own teeth, and feed themselves. I think most people will agree that these are beautiful applications for a well-intentioned technology.

There are other instances, however, where the thinking underlying the use of the technology is more concerning. The proposal to use gene-based technologies to hijack the body's natural immunity from *within* its own cells is an example of what I mean here. Among the concerns regarding this high-tech biohacking that occurred during the Covid-19 pandemic is the now-documented possibility for the artificial instructions injected into the body at one location to migrate through the body to become a part of the genome itself.[8] If this were to occur, the result would be the body perpetuating the cycle of toxin/antitoxin reactions and inflammation that, in theory, would continue indefinitely.

This kind of genetic intervention is an example of precisely why we need to proceed cautiously in applying transhuman technology in real time without fully understanding the consequences of what it means to do so. A closer look into the three phases of transhumanism gives us an idea of what I mean here.

## THE THREE PHASES OF TRANSHUMANISM

The modern transhumanist movement is unfolding as three distinct yet related phases that reflect the current thinking of our society. In light of the rapid growth of technology, it may come as no surprise that all three phases are happening at the same time. It's also interesting to note that there are disagreements within the transhumanist community itself regarding the emerging technologies and how they are used. Not all supporters of the philosophy are equally on board with implementing all of the phases. Following is a summary of each transhuman phase and how it's playing out in our world today.

## Phase 1: Prosthetics and Implants for Enhanced Living

Previously in this chapter, we discovered that, for centuries, humans have successfully used artificial devices that mimic nature to enhance or replace their own biology. Doing so has become so commonly accepted that many people are surprised to discover that the use of these devices is a form of transhumanism. The replacement of limbs lost at birth, in an accident, or in battle, for example, is a familiar application of using artificial replicas to replace natural body parts. The earliest known human prosthetic was discovered in Cairo, Egypt, on the body of a mummified woman believed to be an elite in her society.

When scientists closely examined the woman's body, they discovered that her big toe was missing, possibly from the time of birth, and that it had been replaced by a perfectly and beautifully carved replica toe to give her stability for walking. The toe was sophisticated for the time, being made mostly of wood fashioned to perfectly replicate the size and bone structure of the natural toe. The detail even included an aesthetically accurate nail carved to mirror the curvature of a living toe. The entire appendage was stained to match her skin color and held on to her foot using a system of leather supports and laces. When used, the appendage would have actually worked for the woman and served the purpose of balance that it was designed for.

Modern prosthetics include everything from glasses and contact lenses that mimic the natural ability of the eye to focus, to high-tech robotic knees and hips that are now replacing aging and damaged joints. In recent years, technology has made a new form of prosthetic available in the form of the 3D printing of body parts. A 3D printer, also known as an *additive printer,* uses information from a digital file to lay down repeated layers of a substance fed into the printer, thus building it into a three-dimensional form.

In addition to using 3D printers to create automobile parts, museum replicas of ancient artifacts, and architectural models, it's now possible to have some organs, such as the heart, kidneys, and ears, and even human skin, created from this innovative process. As the technology continues to improve, it can replace the need for painful tissue transplants and the need to have a donor before an organ can be transplanted.

### Prosthetics to Enhance Our Digital Experiences

While we typically think of a prosthetic as an artificial add-on to replace a part of the body that has been lost, the term *prosthetic* actually includes devices that can alter and enhance our relationship to reality and the world around us. A popular new form of prosthetic is the virtual reality (VR) goggle. It allows an individual to have an immersive and sensual experience of sound and images to augment the feeling of an alternate reality.

When the user is engaged in a virtual landscape, the people, situations, and surroundings appear lifelike and are interpreted by the brain as real experiences. While some VR applications can be limited to educational and training scenarios, such as archaeological expeditions and flight simulators, a new level of simulated realities is rapidly gaining in popularity among young people in the gaming community. It's called the *metaverse*, and it's a blending of prosthetics and technology that allows people to engage with one another in realistic ways within the virtual reality. From business deals of mergers and acquisitions to the sale of real estate, remote classrooms, and even rendezvous between partners in one-on-one intimate relationships, all are now possible through the virtual technology of the metaverse.

### The First Brain/Computer Interface

In March 2017, the *Wall Street Journal* published an article describing an example of a futuristic technology that could have been lifted right out of the movie *The Matrix*. The focus of the article was the announcement of a new company founded by SpaceX, PayPal, and Tesla founder Elon Musk—a company that was pushing the boundary between human biology and artificial intelligence.

The company's name was Neuralink, and the product being developed was a specialized computer chip to allow direct communication between the human brain and an external computer, using no wires or cables to facilitate the process.[9] The Neuralink, already being tested on animal subjects, can be implanted through a small, two-millimeter opening cut directly into the top of the user's skull. The device is then placed into the natural space that exists between the brain and the skull, and the bone and skin are then replaced. From the location of the Neuralink on the surface of the brain, 1,024 tiny, threadlike electrodes penetrate downward into the outer layers of the neocortex to communicate with neurons deeper within the brain tissue.

The device itself is a microprocessor that translates the information it receives from the neurons, and then uses wireless technology to communicate those impulses to a typical desktop or laptop computer. In one of the demonstrations, for example, a monkey that had received the implanted chip was taught to play the simple computer game *Pong*. The implanted chip allowed the monkey to make the strategic moves on the computer screen using only thought impulses, without any keyboard interactions.

The goal of the developers is to integrate Neuralink devices with computer-driven prosthetic limbs, such as arms and legs, to empower individuals who have lost limbs to accidents, disease, and war. When he was asked about the viability of the technology, Musk stated that the combination of his company's advanced robotics research with neural-computer interfaces has the potential to build a "robot arm or leg that is as good, maybe better long term than a biological one."[10]

## Phase 2: Human Consciousness/Synthetic Bodies

One of the supposed flaws of our humanness identified by the transhumanist community is our aging process: the deterioration that seems to be a part of getting older, and ultimately, the failure of the organs and systems that make up the human body. Statistically, medical studies show that as we age, for both men and women, there are, in fact, systems of the body that routinely break down as our body chemistry changes and we process nutrients differently. It's this breakdown that typically limits the lifespan of a human body to between 70 and 120 years. I emphasized the word *typical* here, because if we do nothing to support the health, healing, and regeneration of our bodies as we age, then these statistics may be accurate.

However, a growing body of new studies shows that we are born with the innate ability to rejuvenate and regenerate every organ, each gland, and myriad systems of our bodies, and that these abilities are with us until the last breath of our lives. It's because these systems are seldom acknowledged or taught in public schools and medical institutions that transhumanists themselves may not even be aware that they exist. They are steeped in the obsolete belief that the human body must be augmented piece by piece, organ by organ, gland by gland until eventually the entire natural body has been replaced with synthetic approximations.

Advancements of artificial polymers to replace natural skin, computerized sensors to replace our eyes, nose, and ears, and ultimately the invention of an entire synthetic body that could house our natural brain are the goals of this phase of transhumanism. The revolution in modern robotics is rapidly allowing the transhumanist vision of replacing our natural body with synthetic materials to become a reality sooner than we have thought possible in the past.

Until recently, most robots were more utilitarian than aesthetic. I remember sci-fi TV shows in the 1960s and 1970s that portrayed robots that looked like they were made from the familiar stainless-steel beer kegs you'd expect to find at a typical college frat party, with flailing arms made of vacuum-cleaner hoses and cartoonlike features. These make-believe robots served as helpers to science teams exploring new and faraway worlds. Today, in reality, industrial robots that look nothing like humans are used on assembly lines to perform repetitive tasks hundreds or even thousands of times each day, and they do so consistently and effectively.

Manufacturing robots at China's Great Wall Motors automotive plant in the city of Tianjin, for example, perform over 4,000 specialized welding operations with a high level of precision and accuracy.

These kinds of robots have no arms, legs, heads, or eyes. And while we are well into the age of this kind of industrial robot, we are entering a new era of robotics. The new class of humanized social robots that is emerging is gaining acceptance and popularity in ways that rival the futuristic scenarios of popular television series like *Star Trek* and *Battlestar Galactica*.

One of the most advanced, and public, examples of social robots is Sophia, the brainchild of David Hanson and his company, Hanson Robotics. Sophia is an advanced social robot that is made to act, converse, and look human. And she doesn't look like just any human. Hanson intentionally designed Sophia's face to reflect the facial features of two women he admires in his life. Her face is a combination of his wife's features and those of the late British actress Audrey Hepburn.

Sophia made worldwide headlines in 2017 when she passed the citizenship test for the country of Saudi Arabia. Her social intelligence, based upon the OpenCog AI system that is designed for generalized reasoning, made it possible for her to answer the test questions with flying colors. She now has the distinction of being the first robot to be granted full citizenship by a nation. The humanized body of Sophia, which is augmented by AI, is viewed by transhumanists as the technology that will make Phase 2 of their movement possible.

## Phase 3: Consciousness Stored on a Computer Chip

The ultimate goal of transhumanism, according to many of its strongest advocates, is the ability to prolong life, and ultimately, to achieve immortality. In times past, attempts to prolong life centered around the search for an elixir that would restore the physical body to a state of youthfulness and prime health. The search for such a "fountain of youth" is recorded as early as the 5th-century B.C.E. writings of Herodotus and is believed to have led the 16th-century Spanish explorer Ponce de Leon to what is now the state of Florida in the United States, believing that legendary healing waters of rejuvenation existed there. Contemporary engineers are exploring a different approach to immortality.

Some scientists view technology as the modern-day equivalent of the fountain of youth. They believe that recent developments in advanced computer processors now make it possible for an individual's lifetime of thoughts, experiences, memories, and consciousness to be captured and stored on a specialized computer chip. Once this step has been successfully accomplished, the thinking suggests that an individual's consciousness could then be preserved forever and downloaded into a new body whenever an existing one wears out—not unlike replacing the tires on your car or your windshield wipers every few years. This process could, in theory, continue indefinitely, thus accomplishing a form of immortality.

While this process is still in the theoretical stage, it's forcing the scientific community to go back to the basics of biology and answer some of the most fundamental questions of all: What is consciousness? And precisely what is it about our lives that we need to preserve to achieve immortality? So far, the experts have yet to reach a consensus on such questions.

**PURE HUMAN TRUTH 25:** There are three stages of transhumanism, ranging from prosthetic replacement for the body to the proposed capture and storage of an individual's consciousness on a computer chip.

From my personal perspective, there are big problems with the current thinking when it comes to capturing consciousness on a computer chip in pursuit of immortality. Beyond the morality of even attempting such a feat, the entire project is based upon an unproven assumption. That underlying assumption is that human consciousness, and the electrical impulses detected in the brain, are one and the same.

In other words, the idea is that the impulses measured in the brain scan of an EEG are consciousness itself. And because electrical activity can be reduced to the familiar patterns of 1s and 0s that make digital information possible, once this information is captured, it can be stored on a computer and uploaded into multiple and varied forms of life indefinitely.

The problem with this thinking is that it's not supported by the experimental evidence. It's the kind of evidence that I share in the following sections that has opened the door to the deeper questions of what consciousness is and where consciousness lives.

## WHERE DOES CONSCIOUSNESS LIVE?

The Phase 3 transhumanist view is obviously a mechanistic view that assumes that (1) the brain is, in fact, the source of consciousness, and (2) consciousness, memories, and thoughts are the result of chemical and electrical interactions between neurons and synapses. While there is no shortage of theories regarding this phenomenon, neither of these assumptions is supported by the evidence.

To explore this line of inquiry, projects such as Stanford University's Blue Brain project are attempting to build the digital/machine equivalent of a living, biological brain. This project is being carried out on IBM's Blue Gene supercomputer, and has successfully mimicked the relationships of neurons in the brain's cerebral cortex. The thinking underlying this project is that if the biological brain can be duplicated using advanced circuits and computer chips, then the machine should be able to mimic consciousness as well.

An online information series hosted by Stanford University describes the reasoning underlying the Blue Brain project, and others, stating: "According to many neuroscientists, the human mind is really just a complex computer whose function depends on electrochemical processes. In their eyes, if we are able to sufficiently emulate the neural networks that comprise the human brain, it is only natural that intelligence and consciousness should follow."[11] This thinking is a perfect example of how the Newtonian model of the universe actually limits the understandings of life and human potential rather than enhances them.

Seventeenth-century scientist Isaac Newton believed that the universe functions as a vast cosmic machine, driven by moving parts that may be serviced and, when it comes to the human body, replaced when needed. Though advanced for his time, Newton's limited thinking is reflected in present-day attempts to locate consciousness within the body itself. It also illustrates how our willingness to cross the boundaries that have often separated the sciences in the past can lead to powerful new insights, even in scientific disciplines that appear nonrelated.

## THE DANGER OF EMBRACING THE HUMAN-MACHINE CONNECTION

Following the announcement of experiments done to perfect Neuralink technology, the controversy that followed came quickly for two reasons. First, animal-rights organizations responded immediately to the horrors and suffering of the macaque monkeys used in the tests that had computer chips implanted in their brains, and filed legal complaints. The report stated that of the 23 macaques used for the tests, 15 did not survive. In response to this report, Neuralink researchers acknowledged the ethical and philosophical concerns for what those terrible odds would mean if they were to introduce artificial technology into a human body instead of a monkey body, and the proverbial slippery slope it could lead to.

Cognitive psychologist and artificial intelligence expert Susan Schneider, Ph.D., placed into words the kind of concern voiced by academics—scientists and philosophers alike. To merge artificial intelligence with human brains, she said, would be "suicide for the human mind."[12] Schneider's concerns are less about the wireless gaming that was demonstrated by the monkey in Musk's Neuralink laboratory demonstration and more about the door to future applications that this kind of technology has opened. She uses an example from sci-fi writer Greg Egan to illustrate her point.

In his 1995 short-story collection *Axiomatic*, Egan offers a fictional glimpse into a future world where machines are merged with the human body in ways that appear all too real in consideration of the recent advances. In one of the stories, Egan describes the "jewel," a computer chip that is automatically implanted into the brain of all babies at the time of birth.[13] The purpose of the chip is to monitor and record the consciousness of that child's life as a digital backup of their consciousness, their memories, their experiences, and their habits. At some point after they've reached adulthood, their biological brain is removed from their body and destroyed, and it's replaced with the jewel backup chip to perpetuate their life without the deterioration of their brain.

Schneider clearly identifies the problem with this idea. "Because it's implausible to think that your consciousness could magically transfer to the jewel upon the destruction of your brain, it's more likely that at the moment you opted to remove your brain, you inadvertently killed yourself."[14] According to her, the problem with exploring this application for technology is that it suggests "a human merger with AI is ill-conceived—at least, if what is meant by that is the eventual total replacement of the brain with AI components."[15]

Elon Musk is clearly pushing the boundaries of this technology in ways that we've never had to consider in the past. To understand the deep implications of Neuralink, and other examples of merging technology with biology, it's important to understand the biological link between DNA and consciousness.

## CLONES, CONSCIOUSNESS, AND DIVINITY

On July 5, 1996, the Roslin Institute in Scotland, a renowned animal research facility, announced the successful birth of a cloned domestic animal, the now-famous Dolly the Sheep. While Dolly was not the first animal in history to be cloned, she was the first to be cloned from an adult stem cell, known as a *somatic cell*. In Dolly's case, this was a stem cell taken from a mammary gland of an adult sheep. Somatic cells were previously believed to replicate only the part of the body they originated from. This was quite different from the process of using an *embryonic stem cell* that has the potential to develop into any cell in the body.

At first, Dolly's cloning appeared to be a near-perfect success. She looked like other sheep of her species. She lived and behaved like a sheep, and she was healthy enough to mate and produce six lambs during her brief lifespan. At the age of four, however, something unexpected began to happen to Dolly. Her cloned body began to break down. Her health began to deteriorate. The first sign of a problem was a premature form of arthritis that made it difficult for her to walk. For as long as possible, Dolly's pain was managed humanely using drugs that reduced the inflammation in her joints. Within three years, her arthritis had advanced and was further compounded by a progressive respiratory disease that made it difficult for her to breathe. Sadly, on February 14, 2003, Dolly was euthanized.

At the time of her death, Dolly was six and half years of age, having lived only half of the 11- to 12-year lifespan that is typical for members of her species. The obvious question circulating in the scientific community was, "Why?"

What happened to Dolly that brought a premature end to her successful cloning? The honest answer is that no one knows for sure.

While there are theories that include investigating the length of her telomeres, and how her age at death uncannily approximated the age of the sheep that she was originally cloned from, at the time of this writing, scientists still cannot say with certainty why Dolly's body began to break down at a time only halfway through her species' average lifespan. It's this uncertainty that is just one example of a warning flag that should caution us when it comes to engineering the natural processes of life.

We're obviously not sheep, and replacing elements of the human body with machines is also not cloning. There is, however, a common theme that applies to both instances. *For both cloning and transhuman applications, natural processes are being overridden and bypassed* without a full understanding of the implications. It's in this preempting of nature that something appears to be lost in the process. The question is, what is lost?

What is it in the original cells that seems to be lost in their cloned replicas? What element of life was not understood and accounted for in the process of Dolly's cloning? Are we missing the same or a similar element in our own efforts to replace natural blood, neurons, and tissues of the human body with chemicals, machines, 3D printing, and artificial intelligence? And if so, what aspect of ourselves do we stand to lose by replacing our natural cells with artificial sensors, nanogadgets and attempts to capture consciousness on a computer chip?

A report in the medical journal *BMJ* (formerly the *British Medical Journal*) identifies the concerns that have arisen in the scientific community since the highly publicized cloning of Dolly the Sheep in 1996.[16] The article highlights a more recent study of cloned calves, speculating on the possible reason that one of the calves died from health issues only two months after birth. An autopsy of the young calf revealed factors that had contributed to the death, including the discovery that the lymph system, including the spleen, thymus, and lymph nodes, had never developed properly. Citing a study in the prestigious medical journal *Lancet*, the authors infer: "The cloning process seemed to have interfered with the normal genetic functioning of the developing calf."[17]

**PURE HUMAN TRUTH 26:** The inability to clone any form of life that survives its natural lifespan demonstrates that there is something missing in the cloning model, something that is not accounted for in the current thinking when it comes to consciousness and life.

The *BMJ* article uses the *Lancet* study as an example of a very real concern regarding human cloning. "The study could lend weight to warnings that any attempt to clone humans might carry considerable health risks."[18] This warning precisely highlights the concerns that I'm identifying in transhumanist thinking when it comes to replacing some, or all, of our bodies with cloned replicas of our original organs.

It also adds to a growing body of evidence suggesting that something is missing in the current model of what consciousness is, including where it resides. That "something" is the same something missing from the philosophy of Phase 3 transhumanist goals, which are based on the belief that consciousness may be captured as data, stored, and downloaded into successive bodies indefinitely.

The computerized brain simulations, AI beta tests, and cloning experiments are each telling us in their own way that the essence of our existence is not found in the physical expression of our brains or bodies. Yet, transhumanistic philosophy and experimentation are focused exclusively upon the animated portion of our physical bodies. Without accounting for a consciousness that originates from beyond our body, and the need for the biological antennae that create resonance between our body and our consciousness, contact with the essence that drives material life is lost.

To replace portions of the human body with artificial components diminishes our ability to connect with the vast, energetic essence of our existence: our divinity. While an organism may still be alive and animated by the proteins and organs of the body, as we saw with Dolly the Sheep, the consciousness that holds the "program" of that body in place can be lost. This may explain the rapid deterioration and premature deaths of several cloned animals observed by the researchers mentioned in the *Lancet* article.

The mystery of the cloning process is this: The scientists who do it believe they understand the process. They think they've gotten everything correct. They've successfully duplicated the precise genetic code of the organism they're cloning. They've successfully duplicated the environment that allows the organism to thrive. And while it may be true that they've accomplished the preliminary steps necessary for a successful cloning, they've done so without the understanding of how organisms resonate with the intelligent field that programs living systems. This is the organizing field of information that British biochemist Rupert Sheldrake has named the *morphic field* and that he describes eloquently in his groundbreaking research.[19]

It's only now, to take these cloning successes to the next level, that the underlying field of information must be taken into account. A detailed understanding of precisely how the cloning process happens will help to clarify what I mean here.

## CLONING MADE SIMPLE

When we think of cloning, a common image that comes to mind is the removal of cells from the organism to be cloned, such as the calf described earlier, and somehow generating a new calf from the captured cells. And while this visualization is generally correct at a high level, it is incomplete. The actual process is a bit more complex and may be broken into six distinct steps.

With the cloning of a cow in mind, these steps are as follows.

**Step 1.** An unfertilized egg is taken from an adult female cow.

**Step 2.** The nucleus of the egg containing its DNA is removed, *while the mitochondrial DNA (mtDNA) in the fluid outside of the nucleus remains in the cell.*

**Step 3.** An intact cell is taken from the body of a different adult cow.

**Step 4.** The nucleus from this second adult cell is removed and inserted into the unfertilized egg that was taken from the original cow.

**Step 5.** A mild electrical shock stimulates the unfertilized egg, and it begins to grow and divide.

**Step 6.** When the stimulated cell develops and reaches the stage of an embryo, it is implanted into a different cow, where it remains until birth.

From this high-level, six-step overview of a cloning process, we begin to see where the disconnect may occur between an organism and the essence that provides the blueprint for the life of the organism. To do so, once again we need to think of DNA as something more than squishy stuff inside of cells.

For the purposes of our discussion, the DNA of the cloned cells may be thought of as a resonant antenna.[20] Just as it is with any antenna, the DNA receives information from the field of energy that it is naturally tuned to. This is what Rupert Sheldrake previously identified as *morphic resonance*. We can think of the resonant field for the cow as a separate and distinct resonance from that for a human.

During typical conception, a full set of morphic antennae are passed from parent to offspring as the genome of the organism, and the odds of losing access to vital information for the cells are small. As the *Lancet* article suggests, however, cloning the cell appears to disrupt the natural flow of life information within the cells. And now we know why. When we allow ourselves to think of the DNA in terms of resonance and resonant antennae, the disconnect is immediately apparent.

At the conclusion of Step 4 in the cloning description, the process has produced an egg with the mtDNA from one unique adult life while inserting the DNA from the nucleus of another adult life into the empty membrane of the egg. And this is the key.

In the original egg, these two forms of DNA were in resonance and able to send each other correct signals and build the correct proteins, triggering the correct biological sequences to produce a healthy calf. In short, the mtDNA and nuclear DNA were "tuned" to the same information.

For the cloned cell, this tuning cannot happen.

**PURE HUMAN TRUTH 27:** The DNA in the nucleus of a cell must communicate with the DNA that is outside of the nucleus to successfully find resonance with the information that leads to a successful life.

The DNA from different sources, and different lives, are speaking different languages—in other words, they're out of resonance—and therefore are unable to complete their life-affirming communication. This critical detail is vital to understanding the problems with transhumanism. It also explains why human consciousness cannot be captured and preserved on a computer chip.

This understanding is vital for recognizing how altering our natural DNA and replacing our natural bodies with chips in the brain, chemicals in our blood, and sensors under our skin can veil our divinity and divine abilities.

## USE IT OR LOSE IT

There is an axiom in biology that we've all heard, and sometimes use when we joke about the parts of our body that slow down as we get older. The axiom is "use it or lose it." Whether we're talking about geriatric sex or the use of the brain to solve problems, "use it or lose it" reminds us of the simple fact that if we stop moving our bodies regularly, then the systems that allow us to have our experiences will begin to atrophy and eventually stop working.

We often see atrophy of systems when someone is confined to a wheelchair or bedridden for an extended time following an injury. The lack of movement will lead to the muscles losing their tone and elasticity. Eventually a person's strength will wane until it becomes difficult to do simple things that used to be routine, like walk across a room or lift an object without the help of a loved one.

The use-it-or-lose-it principle applies directly to the transhuman replacement of cells, organs, and systems of our bodies with devices that attempt to mimic our natural abilities. Our bodies are driven by the demands of life and the act of movement. When the body senses that it no longer needs to perform a certain task, or to produce a certain response because that response is being performed artificially, our bodies eventually slow down and may even stop performing that particular function. We see a perfect example of this with the discovery of the production of brain cells later in life, and what happens to the cells if they are not used.

For decades, Western medicine held the belief that the number of neurons in the brain is set at birth, and that through lifestyle, environment, and the aging process, we continuously lose brain cells throughout our lives. For much of the 20th century, this belief was the leverage that was used to discourage young people from overindulging in various forms of alcohol. From cottage-crafted beers to vintage wine and traditional whiskey, the story was that with every drink we take, we reduce the number of brain cells in what was already a limited pool of cells to begin with.

A 2012 paper published in the journal *Behavioural Brain Research*, however, gave scientists reason to rethink this belief while opening new possibilities for healing brain-oriented issues like PTSD, depression, and some types of dementia. The paper describes studies conducted on mammals (in this case mice) investigating the growth of new neurons in the part of the brain called the *hippocampus*.

Building upon previous studies showing that neurons continue to be produced throughout life from this part of the brain, the investigation showed that although new brain cells are generated, *they must be used within a short period of time to remain viable*. The paper states: "Most of the cells will die unless the animal engages in some kind of effortful learning experience when the cells are about one week of age."[21]

Although this particular study was done on mice, there is a good reason that scientists often choose to use mice for human brain studies. While a mouse brain is obviously much smaller than a human brain, it is functionally similar: each portion of the brain that performs a specific function in the human brain has a counterpart in the mouse brain. They are so similar, in fact, that the abstract for a paper published by the National Library of Medicine states: "There has been a shift in the proportion of neuroscience-related research using mice from about 20 percent in the 1970s and 1980s to around 50 percent in recent years."[22]

These studies and others point to the phenomenon that is the theme of this section. When we stop using features of our biology such as our memory, stimulating our natural immune response, or the use of our imagination and creativity, these features of our anatomy begin to atrophy. For an individual, in the near future, this may mean the degradation of cognitive function after active neurons are replaced by computer chips or dependence on machines and software to perform the tasks that we used to use our biological brains to do.

**PURE HUMAN TRUTH 28:** One of the dangers of transhumanism is that when we replace our natural biology with artificial technology, the natural functions begin to weaken and atrophy.

Common examples of replacing natural functions with technology include the use of calculators to figure how much money to leave as a tip for our restaurant servers or the amount a cashier at the checkout counter returns in change to a customer who has paid in cash. The frequent and repetitive use of a virtual-reality visor for entertainment, especially by children whose brains have yet to fully develop, may mean the loss of visual sensitivity if the brain is habitually forced into an altered state of perception that is quite different from the reality that the mind is fine-tuned to at birth.

And while these effects, and others, may be temporarily experienced by individuals at one time or another in the courses of their lives, if the effects are consistent, they may be passed to the children of these individuals via genetic factors that determine the next generation.

## BIOLOGICAL EVIL: VEILING OUR DIVINITY

To implement species-altering technology on a widespread basis means that the sensitivities and capabilities that were previously lost to only a limited number of people on occasion now run a greater risk of being lost to entire populations. This includes access to the creative imagination and healing power associated with our divinity. The atrophy and loss of our divine potentials in the current generation, coupled with the potential to transfer this loss to successive generations, is a recipe for a species-wide disaster.

Within one to two generations of accepting transhuman changes into our bodies, the extraordinary potentials of our humanness and its divinity can wither to become vestiges of the abilities that were once common in our past. Eventually they may become a distant memory of what it was once like to be fully human.

To impose this type of change upon entire populations, as is the stated goal that is necessary to achieve the Great Reset, and to do so without the knowledge and consent of those affected, is beyond thoughtless. The consequences of veiling the power of our divinity is more than an unforeseen by-product of progress.

These actions are clear expressions of the malicious force that is driving so much of the pain, war, and suffering that we see in the world today. They are unmasked expressions of the destructive force that must be honestly recognized, identified, and named for what it is. They are expressions of the yet ever-present force of evil. The very definition of evil is: "the condition of causing unnecessary pain and suffering, thus containing a net negative on the world."[23]

With this definition in mind, we see that the global movement to replace our humanness with technology produces a negative effect in the world, as it leads to the suffering that results from the loss of our divinity. In a very real sense, the struggle for technological dominance over our bodies is more than simply a struggle for progress. The battle to rob us of our humanness is a blatant and powerful expression of the deeper spiritual battle between evil and good. This is the ultimate danger of unchecked transhumanism. It is the biological expression of evil.

If transhumanistic technologies are left unopposed and unregulated, to be mandated upon populations across borders and cultures, we stand to lose the characteristics that we cherish as a species. These are the very characteristics that transhumanists view as the flaws of emotional intimacy, empathy, sexual reproduction, and robust immune response. The effort to create these species-wide changes, and to do so intentionally, brings us full circle to the subtitle of this chapter: Erasing the Link between Human and Divine.

The sacred link that allows us to access our divinity and divine power is only possible through our natural, physical, biological bodies. It cannot happen through robotics, 3D-printed organs, and digital 1s and 0s stored on a computer chip. Only our pure human bodies hold the finely tuned DNA that is in resonance with Sheldrake's morphic field and our divinity. To impede our relationship to this field is to oppose the resonant expression of life itself. This is evil made manifest in our daily lives.

There is a good-news "flip side" that may come to a part of the transhumanist movement as well. We may discover that it is only by being on the verge of losing ourselves to technology that we recognize the depth of our natural, extraordinary human capacities.

The Native American traditions of New Mexico's high-desert communities remind us of the truth of our past through a story that they tell to remind themselves of their power as well.

## AN UNEXPECTED MEETING

In 1990, I was hiking along the trail that borders the sandstone cliffs of Chaco Canyon, an ancient and mysterious archaeological site located in the Four Corners area of northern New Mexico. A few weeks later, I would be leading a tour group on an exploration/pilgrimage into the site, and I wanted to check the condition of the trails before asking people to trust me to lead them on the journey. This particular trip would have an unexpected outcome and prove to be a turning point in my life.

As I was hiking into the canyon in one direction, a lone Native American man approached me from the opposite direction. As we got closer to each other, we stopped on the trail to say hello. I shared with him briefly that I was on a reconnaissance trip for the upcoming tour to check out the trail conditions. Having shared why I was there, in return I asked what had brought him to this harsh place, alone, on this particular day.

"I come here to listen to the voices of my ancestors," he said. As he was speaking, he motioned with his hand toward the cliffs on the other side of the canyon, adding, "In those caves."

Only in recent years has the significance of Chaco Canyon been officially acknowledged as the only surviving legacy of a mysterious and forgotten people. While the locals have long recognized the site for its sacred significance, it was in 1987 that Chaco Canyon became a United Nations World Heritage site. And with that designation, there were portions that were closed and made off-limits to preserve them for future generations. The caves that the Native American man was pointing to were among those sites. Without my prompting, the man began to tell the story of his ancestors. Ultimately, what happened to his ancestors long ago in Chaco Canyon gives us deep insight into what is happening to us today.

## A FORGOTTEN TIME

The man began the story by saying that a long time ago, his ancestors lived very differently from the way we live today. There were fewer people to use the resources. And the inhabitants of Chaco Canyon lived closer to the land. They honored themselves. They honored their relationships with one another, and they honored the elements that gave them life. During this time, the people were happy. They were healthy. There was no disease, and the people lived to advanced ages that we can only imagine today. And then something happened.

Although the elders today don't always agree on precisely what that "something" was, the outcome for the stories they share is the same. The people of the Earth began to forget who they were. They began to forget the power they held within themselves. They forgot how to imagine, how to create, how to heal, and how to dream. They forgot their relationship with Mother Earth herself. They became lost, frightened, and lonely. They longed for the connection that they knew was possible, yet they couldn't find it in their lives.

In their loneliness, they began to build machines outside of themselves that reflected the powers they dreamed of. They built machines to enhance their senses of sight. They built devices to amplify the sounds that they could no longer hear and other machines that could send healing into their bodies just the way their bodies used to create healing from within themselves.

The elders say that the story that began long ago continues today. We are the descendants of the lost people. We're still lost today. We still long for the connection with the Earth, one another, and ourselves. The story cannot be finished yet, because it's about us. We're still writing the final chapter of our story.

The elders say that we continue to be lost, frightened, and lonely, and that we continue to build machines outside of ourselves that mimic the powers that lie sleeping within us. They say we will continue to clutter our world and lives with gadgets and devices until the day that we wake up from our dream of longing.

**PURE HUMAN TRUTH 29:** Some Indigenous traditions suggest that we build the complex world of machines and technology outside of us to remind ourselves that they mimic the abilities that already live inside of us.

On that day, we will accept our power and recognize that the complex world we've created around us was us, all along, reminding ourselves of the inner force that we've forgotten. On that day, we will no longer need the machines. We will become so advanced on the inside that our lives will appear simpler to those looking at us from the outside.

In this way, we become in our lives the technology that we have surrounded ourselves with in the world. As we awaken to our true destiny, we dream a new dream.

CHAPTER FOUR

# The Secret

## We Are the Future We've Been Waiting For

The most exciting breakthroughs of the 21st century will not occur because of technology but because of an expanding concept of what it means to be human.

— John Naisbitt (1929–2021),
American author, business consultant

There is an emerging philosophy in the scientific community that has caught the attention of mystics, homemakers, engineers, and young people alike. It mirrors the story in the previous chapter shared by the Native man in Chaco Canyon. The philosophy is simple. The implications of what it means, if it is correct, are enormous. The basic idea of the emerging philosophy can be summarized in a single sentence: *Consciousness informs itself through its creations*. In other words, the things that we design and build in the world around us, are, in fact, us—our collective consciousness—conveying messages from ourselves, about ourselves, back to ourselves.

Through our outer expressions of creativity and imagination, we are asking ourselves to remember something that is important to us—something we need at the time we bring it forward in our creations. This perspective suggests the possibility that what we typically think of as entertainment, including books, fine art, sculptures, music, dance, and film and movies may, in fact, be vital information that we're sending to ourselves to help us navigate the challenges and suffering in the world.

**PURE HUMAN TRUTH 30:** Through our art, technology, books, music, and films, we communicate with ourselves about the things that we are asking ourselves to remember.

Perhaps not coincidentally, this thinking closely parallels the language of writers, visual artists, and musicians when they describe themselves as *channels* for the words that they write, images and shapes that they paint or sculpt, or the melodies that seems to originate from a source that is somewhere beyond them.

Florida-based singer-songwriter Kelly White describes this experience perfectly when she says: "You have to take a deep breath and allow the music to flow *through* you. Revel in it, allow yourself to awe."[1] If we accept the thinking that follows from this philosophy, then the question is, what are our creations saying to us? What are we asking ourselves to remember? The answers may be found in the themes of the creations themselves.

## MESSAGES IN MOVIES

It's been said that there are no accidents in the universe, only unrecognized synchronicities. If this is so, then, in what may be one of the greatest collective synchronicities of modern times, a film was released on March 31, 1999, that set the stage for the thinking that would define an entire technological genre for the new century. The film was *The Matrix,* and its success took the film industry by surprise.

Only nine months following its release, *The Matrix* had created a staggering $466.6 million in worldwide revenue, won four Academy Awards, and earned the distinction of being the greatest box office success of the year. Over two decades later, *The Matrix* franchise remains one of the strongest and most recognizable in the film industry, and the movie's theme is commonly used by philosophers and scientists alike as a metaphor for the topics of virtual reality, human greed, and our relationship to good and evil.

The question that this kind of success triggers is simply: Why?

Why did this science-fiction saga of the age-old battle between good and evil touch so many lives, across so many cultures, religions, and belief systems, and do so on such a deep level? The answer to the question is the doorway to a powerful way of thinking of ourselves and to answering some of the deepest questions of our origins, the nature of heaven, and, ultimately, of God.

The success of *The Matrix* can be attributed to the way it visually engages its viewers. The film revealed a form of never-before-seen action scenes, such as the now-iconic image of the hero, Neo, bending backward 90 degrees while standing to dodge bullets in slow motion and over-the-top martial arts combat taking place in a simulated virtual reality. This kind of action, and the parallels between today's world and the characters' struggle for freedom from the control of an elite few, motivated moviegoers to remain connected with the film.

The plot revolves around the main character's realization that he and his companions are living as digital representations of themselves—avatars—inside the virtual reality of what initially appears to be our familiar, everyday world. Once they become aware of the true nature of their reality, the plot develops into a struggle to free humankind from living as energetic slaves in the digital matrix of the film's title.

## OUR FORGOTTEN POWERS

Beyond the Hollywood shoot-'em-up scenes and the use of computer-generated graphics (CGI), the fundamental message of *The Matrix* is clear. The film is saying to us that there is a world that we can't see from our vantage point in time and space that influences the world we do see, and that we exist in both realities. It's precisely this message, conveyed through familiar forms of entertainment, that has also emerged in the less familiar theories of quantum science. The real-life theories of the universe, such as string theory, which proposes the existence of at least 10 dimensions, along with a growing body of evidence supporting the possibility that we are living in a virtual, computer-simulated reality, make *The Matrix* even more attractive on an intuitive level.

*The Matrix* is not the only modern film that has ignited the possibility of humans having latent abilities that may be harnessed to benefit our everyday lives. Following the success of *The Matrix*, a new genre of film emerged that used similar methods, including similar CGI tools, to convey stories that express powerful themes that question the nature of our reality.

On December 18, 2009, the first installment of the *Avatar* film series was released. The multiple-award-winning film was immediately embraced by a worldwide audience and became the highest-grossing film of all time, a distinction that it held for nearly 10 years following its initial release. In a theme similar to that of *The Matrix*, *Avatar* revolves around human consciousness being projected into another reality by people animating their avatar equivalents so they can interact with beings living on a faraway planet.

The tall, graceful beings known as Na'vi that inhabit the planet where *Avatar* takes place hold a deep reverence for nature and the natural forces of their world that are used to heal and evolve. It's precisely this pristine world of nature that the human explorers want to ravage, just as they've done to the point of depletion back home on Earth. It's this stark contrast in thinking between integrating with nature and plundering from nature that provides the plot for the film.

As the leading human avatar becomes immersed in the beauty and the sacredness of the world that he's been tasked to infiltrate and exploit, his relationship and respect for the Na'vi deepens.

From the perspective of consciousness informing itself through its creations, *Avatar* reminds us of the sacredness of our own natural landscapes, as well as the dwindling numbers of Indigenous inhabitants surviving today who are engaged in a struggle to preserve and protect the resources and native beauty of our own world.

The messages in our present-day films are not limited only to avatars and virtual realities. The huge acceptance of superhero films like *Wonder Woman*, *The Avengers*, and *Mortal Kombat 2021* each offer insight into a deep sense that there is more to us than we've been led to believe. These films and others in the same action-fantasy genre highlight individuals with dormant superpowers that are awakened precisely at the time when the world needs them most. These movies all tell us we're not what we've been told, and probably even more than we've allowed ourselves to believe. Each one delivers a message of personal empowerment and the ability to successfully transcend the challenges that the world brings to our doorstep.

**PURE HUMAN TRUTH 31:** The most popular and successful movies of our day are those that portray humans remembering or discovering hidden talents and superpowers.

If the philosophy that we are communicating with ourselves through our creations applies to the arts, then it makes sense that this theme applies to other creations as well. This includes creations made with technology. What is the advanced technology that is birthed in places like America's Silicon Valley and its high-tech Asian counterpart, Singapore, saying to us about ourselves? What latent abilities could the Internet, AI, and superfast computer chips possibly be reminding us of within ourselves? And what does the rapid advancement of technology tell us about the timing of our own awakening to new potentials within ourselves?

## WHY NOW?

The 1960s saw modern technology transition from the use of clunky, surface-mounted electrical components and rarified-gas-filled vacuum tubes to electrical circuits imprinted on microchips that are so tiny we need a magnifier to see them clearly. In 1965, Gordon Moore, the co-founder of Intel Corporation, recognized that the number of components on a microchip was doubling approximately every two years. It was immediately obvious to him that, at the rate of doubling he was observing, some kind of limit would be reached by the 2020s, if not sooner. The fact that the materials from which chips and circuits are made contain atoms, and that atoms by definition have fixed properties, suggested to Moore that the properties of the elements themselves soon would set limits on their functioning.

During an interview in the early 2000s, Moore stated, "The fact that materials are made of atoms is the fundamental limitation and it's not that far away . . . We're pushing up against some fairly fundamental limits so one of these days we're going to have to stop making things smaller."[2] The limits that Moore envisioned are being reached just as he predicted.

In 2012, Intel announced the first 22 nanometer (nm) processor, followed by a 14 nm chip in 2014 and a 7 nm chip in 2024. For context, one nm is one billionth of a meter. Stated another way, a single human hair is about 90 nm wide, so a 7 nm processor is about 1/12th of the diameter of a human hair. These comparisons help to understand why the technology is limited in the way that it is. The information can only flow as fast as the energy can move between the atoms of the chip.

Intel has recently revealed that they now have the technology to produce a chip that is 2 nm, bringing it very close to the ultimate limit that Moore predicted back in 1965. If the technology in the world around us is reminding us of something within us, what are these extraordinary processing speeds, and their limitations, asking us to remember?

## PURE HUMAN TECHNOLOGY

In the previous chapter, we considered the DNA in the nucleus of our cells from an IT perspective. In doing so, we were able to recognize the potential of the double helix as an efficient storage medium for huge amounts of information. When we expand this thinking to include the entire cell rather than just the nucleus, we gain new insights into what it is that consciousness is nudging us to remember. This perspective reveals an astounding possibility that bridges movies like *The Matrix* with the mysterious and mythological stories that our ancestors told of superhumans from our distant past. And it's based upon new, peer-reviewed discoveries made using the best science of the modern world.

In the following sections of this chapter, I'll highlight some of the compelling discoveries I find most inspiring to provide a better sense of what I mean. The purpose of these sections is to offer a reason to think of ourselves differently than we have in the past—as more than animated bodies of tissue and fluids. Rather, we'll discover that we are a highly advanced, technologically sophisticated technology—*a soft technology*—with capabilities that far exceed anything that could be produced in our modern laboratories.

**PURE HUMAN TRUTH 32:** We are a sophisticated soft technology with the functions of our cells meeting and, in some cases, exceeding the capability of the functions of AI and the components on a computer chip.

Let's begin this survey of our soft technology with a look at several fascinating properties of the body itself. Scientists tell us that the average human body consists of somewhere in the neighborhood of 50 trillion individual cells. Each of these cells, in turn, tells a powerful and perhaps unexpected story of untapped potential and superhuman capabilities.

## Property 1. Our Cells Are Electrical

Each human cell generates an electrical voltage of about .07 volts. Although this is, admittedly, a small voltage, when we consider the number of cells that are simultaneously producing this voltage, the math tells the story. Multiplying the 50 trillion cells of our body by the .07 volts each cell emits yields a mind-blowing 3.5 trillion volts of electrical potential for a single human body! This is such a staggeringly large voltage that it can be hard to wrap our minds around what this number represents. The following analogy may help.

Let's consider a familiar 12-volt automobile battery to help us fathom what the potential power in our bodies really means. In other words, if we lined up rows of 12-volt automobile batteries, side by side, it would take three billion of them to match the electrical potential within one human body. It would take 8,078 of these batteries to fill a single football field, and it would take 371,000 of these football fields, each filled with 12-volt batteries lined up, end to end and side by side, to equal the inherent electrical potential within each of us.

Technological comparisons don't stop with only the combined electrical potential of cells. Each cell of the human body also functions naturally as a tiny electrical circuit, with the biological equivalents of capacitors and resistors regulating the energy that passes through it.

The first time I saw this analogy of a human cell as an electrical circuit was in a declassified research paper that had been published in the USSR during the Cold War years of the 1980s. While research in the United States lagged years behind the perspective of the USSR at the time, and the American scientific community had been reluctant to think of human biology from a physics perspective, this was not the case in other countries. Asian, Russian, and Indian scientists were publishing, and continue to regularly publish, research that gives us deep insights into our physiological healing processes and our abilities to self-regulate our extraordinary potentials.

A paper published in 2024 in the *Bangladesh Journal of Medical Physics*, for example, parallels the study that I saw from the USSR mentioned previously, and describes the biological components of the cell as the electrical equivalents of capacitors and resistors as shown in the illustration below.[3]

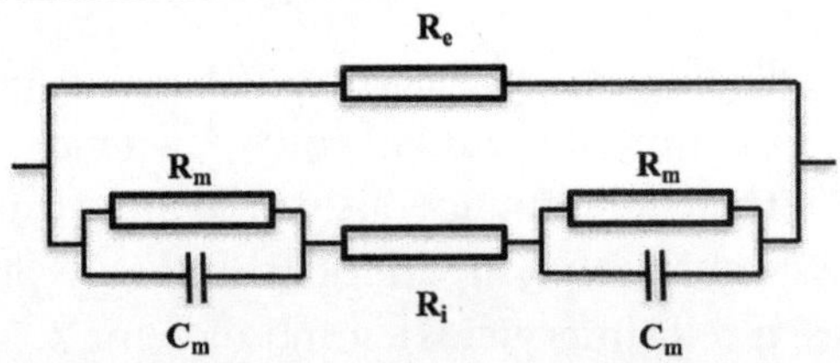


A schematic diagram of the electrical circuit formed as the biology of a single human cell. In the drawing, $R_m$ and $C_m$ indicate impedance for the cellular membrane. C indicates a capacitor, R indicates a resistor, and the sub m indicates that those properties are describing the membrane of the cell. Adapted from the *Bangladesh Journal of Medical Physics*.

The *Bangladesh Journal of Medical Physics* paper describes how embracing the electrical nature of the human cell opens the door to novel and innovative healthcare applications, such as the early detection of breast cancer, in countries where the scanning devices needed to do mammograms is scarce and access is limited.

## Property 2. Our Cells Are Transistors, Resistors, and Capacitors

To regulate the electrical energy in our bodies, each human cell functions as a transistor—a miniature semiconductor that regulates and amplifies electrical energy in a circuit. Each cell also functions as a resistor, both regulating the flow of energy through the cell and providing a specific voltage for the transistors. Each cell functions as a capacitor, storing and releasing energy at a specific rate and time. The very mention of such components sounds like a throwback to the electronics that we used to see when we opened the backs of our televisions, radios, clocks, and guitar amplifiers in the 1960s and 1970s.

When I mention these components of household technology in today's seminars with younger audience members, they often look perplexed at the sight of the images I've projected onto the screen. To younger generations, the images look like foreign objects from another world that have no relevance in the modern day. That is, until I flash the image of a miniaturized printed circuit on the screen and tell them that all of the functions performed by those large and clunky components of the past can now be contained on sleek, sophisticated, tiny microchips. Then they nod their heads in understanding, as the chips are all that they've known in their lifetimes, and they can easily relate to them.

But our survey of our inner soft technology doesn't end here.

## Property 3. Our Cells Absorb and Emit Light

Each cell of our bodies absorbs and releases photons, the quantum stuff that light (and ultimately every atom) is made of. *Photons* are elementary particles of quantum energy that travel at the speed of light. Beyond the metaphor often used in New Thought communities that we are "beings of light," we literally generate photons from within our cells, and this light we generate can be measured and photographed with advanced scanners designed to do so.

But as interesting as that fact is, the light that we absorb and transmit is more than just a bright flash of energy. Light is information. The machinery in our cells absorbs wavelengths of light that we receive from the world around us, and translates those wavelengths into meaningful instructions that feed our vital functions. These wavelengths of light give us life and promote our healing.

It's only recently that the scientific community has acknowledged the existence of a field of energy associated with each human body—a *biofield*—and the role that accessing our biofield can play in maintaining our health and healing from injury and illness.

The abstract for a 2015 publication published by the National Center for Biotechnology Information describes the value and potential of exploring the human biofield. It reads: "Biofield science is an emerging field of study that aims to provide a scientific foundation for understanding the complex homeodynamic regulation of living systems. By furthering our scientific knowledge of the biofield, we arrive at a better understanding of the foundations of biology as well as the phenomena that have been described as 'energy medicine.'"[4]

In case you're wondering, the term *homeodynamic regulation* is a more nuanced idea than the idea of *homeostasis*, which is the concept that the body always tries to restore itself to its optimal state of temperature, hydration, nourishment, and oxygenation, among other things. That's why we sleep, eat, and drink water and shy away from directly experiencing extremes of cold and heat, for example. Homeodynamics is a more diverse form of autoregulation. It provides stability to our bodies, but leaves room to accommodate adaptation and growth.

As the science of the human biofield receives greater acceptance, we can expect doors to be opened for remarkable innovations in healthcare. These will emerge from a perspective that is distinctly different from the chemically oriented model that permeates the field of medicine today.

For example, we have long used scanning devices to detect tumors in the human body. The scans we have detect various forms of energy, ranging from heat images detected by thermal devices to radio signals detected by magnetic-resonance imaging devices. But although we are comfortable using energy to *detect* the anomalies in the body, we still typically use the physical instrument of a scalpel or a laser beam to *remove* anomalous tissue once it has been discovered.

Biofield technology opens the door to the possibility of using energy to actually heal the tissue rather than removing it. And if the need to remove certain amounts of tissue persists, biofield energy could be "tuned" to eliminate said anomalous tissue without the need to cut into the body and expose it to the risks of infection and complications of surgery.

This advanced technology, and more like it, becomes possible as we allow ourselves to cross the traditional boundaries that have in the past separated sciences such as medicine, chemistry, and physics from one another.

## Property 4. Our DNA Is a Soft Antenna

In Chapter Three, we discovered that the long molecules of DNA in the nucleus of a cell function together as a biological antenna. Such antennae find resonance with specific signals from both our physical and our emotional environments. The smaller segments of DNA that make up the genes function in the same way, and are tuned to even more specific types of information that provide even more specific signals, in the process creating the blueprints for health and healing in our bodies.

Once again, the term *antenna* is more than a metaphor here. As biofield research reaches beyond pure biology into related areas of study, scientific research is telling us that we are intimately enmeshed with the energetics of our environment in ways that we're only beginning to understand. A 2017 paper that was published as part of the proceedings of the 2016 International Conference on Soft Computing: Theories and Applications states, "We report that 3D-A-DNA structure behaves as a fractal antenna, which can interact with the electromagnetic fields over a wide range of frequencies. Using the lattice details of human DNA, we have modeled radiation of DNA as a helical antenna."[5]

In technical terms, this paper reiterates the principle that is the vital link between our bodies and our divinity, the fact that the molecules of life function as antennae in our cells.

The authors of this paper have published additional work describing how it is due to the antenna-like characteristics of DNA that certain frequencies in the environment, both natural and human-made, may play a role in both the formation, as well as the healing, of certain diseases, including some cancers.[6]

Interestingly, it's this antenna-like function, which fosters communication between the DNA located within and outside of a cell's nucleus, that is required for the successful cloning of a life-form like Dolly the Sheep.

### Property 5. Our Cells Store Information

As described previously, the DNA within each cell in our bodies can function as a lasting repository of information. What I did not mention before, however, is that there are two sources for the information that may be stored in the DNA.

First, it's possible to store the digitized version of written information into the structure of the DNA, and then retrieve it at a later time. The examples cited in the research explored in Chapter Two come from studies where the DNA holding the digital information was passed through successive generations in an organism before the information was retrieved, still intact and readable.

The second kind of DNA storage is not from information sourced outside of the body. Rather, this information comes from within the body itself. Each strand of DNA holds a record of every successful genetic mutation that has ever occurred, at any time, both for an individual and ultimately the transactions that have been embraced in the history of our species. In fact:

- Our DNA record is permanent. As long as one human remains alive anywhere in the universe, within the 50 trillion cells of that human, there is a permanent record of the genetic history of our species.
- Our DNA record is transparent. It is not cloaked behind a hidden biological "firewall" or tucked away in a place where no one will see it. The record of our genetic history is transparent and available for all who have the skill and technology to access and read genetic code.
- Our DNA record is secure. The DNA transactions from our past are with us forever. All species-wide DNA transactions are coded into each of the approximately 2.5 billion cells that are produced each hour in our bodies.

These principles of permanence, transparency, and security are so efficient, and so sophisticated, that they are mimicked today in an emerging technology that is revolutionizing the world of money, banking, and decentralized finance. This *blockchain technology* was the foundation for the first and most successful application of it yet, which is the cryptocurrency Bitcoin.

**PURE HUMAN TRUTH 33:** In addition to our cells functioning like electrical components, our DNA preserves a record of the genetic transactions that make our species as it is, a blockchain-like record that is transparent, permanent, and immutable.

When we consider the soft technology of the human body from an IT perspective, it is immediately apparent that we are an extraordinary form of life, and our existence is the result of much more than lucky biology and random mutations. The cells of our body perform functions that software engineers, hardware engineers, and systems designers dream of creating in the world of synthetics and robotics. Our cells self-replicate, self-diagnose, self-heal, and self-select in a way that the best engineers at work today, using the most advanced technology available, fall short of accomplishing.

Additionally, when they are damaged beyond repair, human cells self-eliminate to ensure that partial or incorrect DNA information is not passed to new cells through cell division. The process of *apoptosis*, or programmed cell death, is a vital part of the healing dynamics of the body. It is estimated that our bodies eliminate between 50 and 80 billion cells each day through apoptosis, while healthy cells continue to divide and replicate themselves. It is this cycle of death and renewal that maintains the delicate and constant balance of the cells in our bodies that give us life and healing.

## HUMAN NEURONS VS. COMPUTER CHIPS

Among the most mysterious of the cells that make up our bodies are the neurons. They're mysterious because we're still discovering how they're made, how they function, and what they are capable of. While we commonly associate neurons with the brain, and it is true that the greatest concentrations of our neurons are located within our brains, they're also found throughout the rest of our bodies.

The brain and spinal cord form the body's *central nervous system* (CNS) and are estimated to contain approximately 86 billion neurons. Beyond the CNS, the body's *peripheral nervous system* (PNS) contains the neurons that account for the nervous system in the rest of the body. The exact numbers of these nerve cells is unknown, but the figure is estimated to be in the neighborhood of another 25 to 40 billion.

As different as the roles for the various neurons are in the body, the cells themselves all have a similar form and function in a similar way. Neurons communicate with one another in order to share information using electrical signals, generated as a result of our thoughts, perceptions, and beliefs, as well as chemical signals that move across the gaps between cells, the *synapses*. It is the ability of either the chemical or electrical information to cross a synapse successfully that determines the level of our cognition and cognitive abilities.

Neurons play a vital role in our soft technology. In some ways, natural neurons are actually superior to, and even outperform, computer chips designed for similar purposes. For example, to date, they work faster.

The speed of a computer's processor is determined by its internal clock speed. The internal clock generates the electrical pulses that regulate the functions on the chip. When the first generation of computers was built in the mid-20th century, their clock speed was initially measured in terms of kilohertz (kHz). One kHz represents 1,000 cycles per second. Only a few years later, however, in the 1980s, the clock speeds had increased so quickly that the units of measure were changed to millions of cycles per second, or megahertz (MHz).

Now, in the early years of the 21st century, clock speeds are measured in the familiar gigahertz or billions of cycles per second (GHz).

But the speed of computers may soon increase to match the speed of the human brain. As quantum computers become available, the concept of a clock to limit processing speeds will become obsolete as the computers' processing power will be directly linked to the fundamental forces of the universe and the speed of light. These are precisely the forces that already form the foundation of our natural neurons.

While such advances in computing power will definitely open the door to a new era of superfast quantum processing, they do not necessarily mean that the capabilities of artificial computers will surpass those of the natural human brain. The reason is that natural human cognition will continue to scale its abilities at the same time that the upper limit on computer processing is expanding. As we discovered in Chapter Three, the upper limits and capabilities of the human brain, and in particular, the states of consciousness it is able to achieve, now appear to be scalable beyond any limits known today.

While transhumanists often compare the function of the human brain to the function of a computer chip in an attempt to illustrate the superiority of computer technology, there are features of neurons and the natural brain that are superior to the features of a synthetic chip and a computer. In addition to the processing speed of the human brain not being limited to the speed of an internal clock, some scientists, such as neurosurgeon and neuropsychologist Karl Pribram (1919–2015) and his colleague theoretical physicist David Bohm (1917–1992), suggested that the information conveyed throughout the brain is holographic.[7] This means that the data is not confined to the physical conduits of the neurons themselves.

Recent studies suggest that our brains may be so advanced, in fact, that they process information using relativistic principles that were first proposed by Albert Einstein, with information traveling across the neural networks at the speed of light.[8] Beyond these fundamental differences in the way information is transmitted, the sheer processing speeds of the most advanced computer chips and of a natural neuron are surprisingly similar.

As described previously, the bulk of the processing on a computer chip is accomplished using transistors that regulate the flow of information. Because the synapses between neurons perform a similar function in the brain, for the purpose of this comparison, it makes sense to consider these two processes as rough equivalents.

A study conducted by the University of California at San Francisco and the Salk Institute summarizes these parameters, reporting that "a modern microprocessor chip has $10^9$ transistors, the human brain contains about $10^{14}$ synapses (and a brain uses about as much power as a microprocessor)."[9]

To accomplish the next level of actually calculating the relative processing speeds of a brain and a microprocessor, the researchers chose to base their calculations on the generally accepted facts that there are an estimated $10^{10}$ neurons in the human neocortex, and that the neurons generally fire at a speed of 10 Hz. This means that the brain is processing about $10^{11}$ instructions per second. The paper concludes that calculations performed by the human brain would yield "100 fold more than the $10^3$ million instructions per second (MIPS) of a modern microprocessor with multiple cores (computing units)."[10] In other words, based upon these calculations, the artificial computer is about 100 times slower than a natural human brain.

**PURE HUMAN TRUTH 34:** The capacity and function of a human neuron is adaptable and scalable for performance beyond any known limits rather than being limited in the way that physics limits the capacity of solid-state computer chips.

Another fact to take into consideration is that the brain performs a kind of "information triage" when it receives input, such as it would while doing facial recognition or discerning a threatening situation. The information is segmented and processed in parallel functions, yet at different speeds, depending on which part of the brain is doing the processing. The amygdala of our brain, for example, is the hub for processing fear, while the hypothalamus is where we process the smells, touch, and sounds that we associate with love. This feature is in stark contrast to the serial processing of a computer. Although a computer is really fast, it does not associate disparate pieces of information in the parallel fashion that we humans are capable of.

The biometric systems that are becoming common at airport security checkpoints use a form of serial processing based upon facial recognition. As we approach a security kiosk in an airport, overhead sensors scan our faces and a digital record is made of our facial contours and of features such as the distance between our eyes and the space between the tip of our nose and the top of our lips, among others. That information is then compared to a digital record of hundreds of thousands or even millions of other people in order to find what is hopefully a good match. This way authorities can discern that we aren't a threat to the public or detain us if they suspect we committed a crime—or might.

Studies show that while humans can recognize a familiar face in as little as 360 to 390 milliseconds (one millisecond is one-thousandth of a second), typical computer-based facial recognition systems can take as long as one to two full seconds to perform their tasks.[11] And while computers have been found to be more accurate for simple still images, humans still have the edge when it comes to recognizing complex facial scans or moving images.

## WHY OUR HUMANNESS PREVAILS OVER TECHNOLOGY

In addition to the remarkable human capabilities described in the previous sections, we humans also have another ability that gives us a very special edge. It is our extraordinary ability to *self-regulate* many of the potentials that lie dormant within us. We are the only known form of life, for example, that can choose to create a stronger immune response in our bodies; that can choose to create greater resilience to the big changes in our lives and in our world; that can choose to relieve stress all the way down to the molecular level of our DNA that awakens deep healing, and more. The key here is that we are the only creatures on Earth that can initiate such super potentials on demand, in the place and at the time of our choosing, rather than waiting for cues from the world around us to awaken these states of being.

The beauty of our human design is that we don't need to know the science of how these systems function in order to experience their benefits.

The way we are created is so sophisticated that gaining access to the complex interactions happening behind the scenes in our bodies is made easy through the "user interface" of our perceptions. Our everyday experiences of thought, feeling, emotion, belief, breath, and focus are the triggers that can set into motion the cascade of chemical reactions that we're typically unaware of. Yet these are the very events that can fill our lives with health, healing, and ease, or conversely with disease, suffering, and despair.

**PURE HUMAN TRUTH 35:** We self-regulate our advanced soft technology through the user interfaces of thought, feeling, emotion, breath, and other epigenetic factors.

One of the telltale signs of a really advanced technology is that the user interface—the dashboard to access the functions of a device—is simple and easy to use. For example, when we switch on the smartphone that we hold in the palm of our hand, we activate apps and information channels by simply tapping the screen with our fingertips, speaking to a programmed virtual assistant, or selecting highlighted fields. We can accomplish entire tasks without ever actually typing a single word of instruction.

We typically have no idea of what is happening behind the scenes that makes it possible to connect us with our friends and family or to transfer money from our bank account to pay a bill. And we don't need to know. That's the beauty of the user interface. The layers of hardware, firmware, and software between us and the functions of the machine do the work for us.

Our human "user interface" to access our extraordinary technology works in much the same way. It's through engaging in different combinations of familiar, everyday actions that we access, trigger, and activate an impressive range of potentials available to us within our bodies.

## THE DISCOVERY THAT SENT SHOCK WAVES THROUGH THE SCIENTIFIC COMMUNITY

On June 26, 2000, then president Bill Clinton did something that very few U.S. presidents have ever done during their term in office. On live television, speaking from a podium in the East Room of the White House, he announced the completion of a scientific project, one that had been years in the making: the mapping of the human genome. Comparing the Human Genome Project (HGP) to the project of mapping the United States of America 200 years earlier, Clinton stated that it was of even greater significance. "Without a doubt, this is the most important, most wondrous map ever produced by humankind," he said.[12]

While the mapping of the human genome was an immense achievement to be celebrated, some scientists were more mystified by what the map revealed than they were in awe of the accomplishment. The reason is that the results of the program overturned a longstanding theory regarding DNA, genes, and the role genes play in the body.

Prior to the announcement, there had been a widely accepted belief that each gene of our DNA programs only one specific protein in the body. This perceived one-to-one correspondence led to the thinking that if we could isolate the genes responsible for specific proteins, then we could use therapies and create drugs to modify any gene not functioning properly.

The human body has more than 100,000 unique proteins that work together to support all of the organs, glands, tissues, and systems in our bodies. Before the Human Genome Project, the expectation was that, at the rate of one gene producing one protein, there should be at least 100,000 genes to be discovered and mapped. Eventually, the expectation was, the success of the gene map would lead to the creation of 100,000 new drugs—one to address the functioning of each of the expected genes.

Midway through the project, however, it became obvious that something was wrong with this hypothesis. And not wrong by just a little bit. Hugely wrong. The number of genes mapped though the midpoint of the project was so far below what had been projected that scientists were both shocked and confused.

By the end of the project, a new human story had been revealed. Our entire genome turned out to be the result of only about 24,000 genes. And the discrepancy between expectation and reality was not the result of a mistake in a math formula or a simple rounding error. The map had accurately revealed that we have only about 25 percent of the genes that the theory of the era had predicted.

There was only one logical explanation for the huge discrepancy. The existing theory about genes and proteins matching each other was wrong.

**PURE HUMAN TRUTH 36:** In the year 2000, scientists discovered that the human genome is made of only about 24,000 genes, meaning that the previously accepted relationship of one gene producing each of the roughly 100,000 proteins in the human body was wrong.

The discovery that so few genes are managing the complex systems of the human body was like an earthquake that sent shock waves throughout the scientific, medical, and pharmaceutical industries. Due to the complexity of the human body when compared to other forms of life, the expectation was that we must have a lot of genes to run our complex biological machinery. The mindblower was that "lesser" forms of life had a comparable, or even larger, number of genes than we do!

A simple field mouse, for example, has 25,000 genes—1,000 more than we complex humans—while a flowering lotus plant is now believed to have in the neighborhood of 15,000 more genes than we do, with an estimated 40,000 genes.

Clearly, something was off with the way genes had been considered in the past. The genetics that had been taught in classrooms, used to justify millions of dollars in research projects, and touted as the basis for promising new drugs was wrong. And that "something" is what opened the door to the radical discovery that highlights why we have the powers that we do, and the role of our divinity and divine processes.

## EPIGENETICS: THE DISCOVERY OF MULTIFUNCTION GENES

The Human Genome Project revealed that our genes are not "fixed" when it comes to their function. They aren't static segments of DNA that perform one, and only one, task.

Rather, our genes are dynamic. Meaning, they're malleable and they program themselves based upon the demands of the environments in which they find themselves. Subsequent studies have shown that a single human gene can program as many as 100 different proteins, a discovery that explains why the 24,000 genes in our bodies can produce up to 400,000 unique proteins, including the 100,000 or so that keep us alive.

Such discoveries have led to the emergence of a "new" science that is called *epigenetics*. The prefix *epi* means "above," and *genetics* refers to the genome. It follows, then, that the study of epigenetics investigates factors above or beyond the genome itself that trigger specific genes to produce specific proteins.

I'll describe some of these factors in detail in the following sections.

Above, I placed the word *new* in quotes when describing epigenetics. The reason for this is that while the term *epigenetics* has become popular in recent years, the theory itself isn't new and was actually developed in the late 1960s. It was based upon laboratory studies that revealed the power of the environment to program genes.

The principle of epigenetics is often demonstrated using the simple experiment of cells in a petri dish, as in the following example.

We'll begin our experiment with a group of identical cells cultured in a typical laboratory petri dish. The cells are from the same source, are of the same age, and have identical properties. If we take some of the cells from the original petri dish and place them into another container, and this container is exposed to a different environment, the cells begin to develop differently. The trigger for change can be a shift in the chemical environment or the nutritional environment, or the result of a shift in the emotional environment. This is the power of the genes.

A single gene can produce many different proteins. This is possible because different segments of the genes (called *exons*) can combine in different ways to build different proteins. A cell exposed to one environment can become a muscle cell, for example. Another cell, taken from the same original petri dish of identical cells, then exposed to a certain environment, might become a blood cell, a liver cell, or a bone cell, or another type of cell. The principle of epigenetics explains why this is possible. Different environments awaken different segments of the genes so that they can respond to the environment in a way that reflects the environment.

**PURE HUMAN TRUTH 37:** Through epigenetics, it is possible for one human gene to "program" up to 100 unique proteins. This means we can do more programming with a few highly efficient genes.

When we see the principle of neurons that change, grow, "wire" together, and "fire" in response to the emotional environment we provide, we call the process *neuroplasticity*. The plasticity of these nerve and brain cells tells us that the nervous systems we're dealing with are not set into stone. They're not static and rigid. Rather, they are, as the term *plastic* suggests, malleable and dynamic. This principle applies to our genome as well. It is the *genetic plasticity* of our DNA that empowers us to free ourselves from the hurt, suffering, and trauma that may be programmed into our genome through epigenetics and epigenetic factors.

There are a variety of epigenetic triggers, and any one of these factors, or a combination, may be responsible for the DNA configurations that bring us continued life, joy, and healing, as well as suffering, fear, and disease. The epigenetic triggers can be environmental factors, such as toxins in the air we breathe and the food we eat. They can be substances that we ingest, such as prescription pharmaceuticals and recreational drugs, including both synthetic and natural plant-based psychedelics and alcohol.

Perhaps the most potent of the epigenetic factors that influence every facet of our lives, however, are factors that cannot be seen and aren't easily measured. These factors come from the energy of our conscious and subconscious emotions, our thoughts, and what we believe to be true about ourselves. Although we have personal access to these invisible factors, they're sometimes the most mysterious and challenging to shift.

Discovering our ability to shift our beliefs is where we find our greatest mastery in life. When we choose to upgrade our life story and think differently about ourselves, epigenetic principles tell us how the shift in our perception actually changes our bodies.

The discovery of mirror neurons is a perfect example of how this process works.

## AN UNEXPECTED DISCOVERY IN THE BRAIN

In 2004, a group of Italian scientists announced a discovery with implications that continue to rock the worlds of psychology, biology, and evolution. The report of the research that led to this discovery, published in the peer-reviewed journal *The Annual Review of Neuroscience*, announced an evolutionary "missing link" providing a previously unrecognized pathway to advanced states of cognition and learning. The article identified a new kind of neuron in the human brain called a *cubelli neuron*, or more commonly, a *mirror neuron*.[13]

Mirror neurons are located in portions of the brain that relate to key functions like touch, temperature, and pain. This is important because these functions are located in the most recent area of the human brain to develop, the one that gave us our humanness when we first appeared 200,000 or more years ago. What sets the mirror neurons apart from other neurons of the brain is the way that they function. Experiments reveal that the mirror neurons are triggered, or fire, under either of two very different sets of conditions.

First, not surprisingly, a mirror neuron will fire when we are having a lived physical experience. Whether we are hiking alone, immersed in the beauty and cool, thin air of a high-altitude mountain trail or we are exploring the sensual intimacy of physical contact with a romantic partner, our mirror neurons are activated and respond to this lived experience by sending signals to the parts of our brain and body that release chemicals to support us in our experience.

During a romantic encounter, for example, we would produce bonding chemicals such as dopamine and oxytocin, as well as adrenaline in response to the excitement and the imagined possibilities that we sense in the moment. Additionally, males may produce the hormone vasopressin to strengthen the bonding experience by identifying their sexual partner as a preferred partner.

The key information here is that our body chemistry changes in response to an experience that we are actually living in the moment.

It's the second trigger condition that is perhaps most surprising and mysterious. This trigger holds the greatest potential when it comes to our capacity for self-mastery, healing, and regulation of our bodies.

The 2004 discovery revealed that our mirror neurons also fire when we simply watch someone else having an experience. In other words, from the perspective of the neurons, they don't know the difference between having an experience and witnessing an experience.

**PURE HUMAN TRUTH 38:** Mirror neurons do not know the difference between having a lived experience and watching someone else have an experience.

The implications of this discovery are vast and controversial, and they particularly run deep when it comes to personal responsibility and mastery. Additional experiments have opened doors to possibilities that begin to sound like the stuff of science fiction. Except these possibilities aren't fictional; they're very real. And once we see them, we cannot unsee them.

It is up to us to discount or embrace what they mean for our own lives.

## OBSERVED VS. IMAGINED: THE BRAIN DOESN'T KNOW THE DIFFERENCE

Brain scans documenting an experiment done at the prestigious Salk Institute in La Jolla, California, published in 2007, illustrate the ways that our mirror neurons fire.[14] This experiment consisted of using scanning technology to document the activity of mirror neurons in the brain of one individual while that person was performing three different tasks. The first line of the scan shows the parts of the brain that were activated by the simple act of the subject reaching for an object. Not surprisingly, the portions of the brain that "lit up" were the places commonly associated with the functions of extending the arm and opening the hand to grasp a physical object.

In the second part of the experiment, the subject being tested did not physically reach for anything. Rather, the individual simply *observed* another person who was actually doing the reaching. The second line of the scan reveals the power and the mystery of the mirror neurons. It shows that the same locations in the subject's brain were lighting up as if they were actually doing the lifting. While the intensity of the activity was slightly diminished, for all intents and purposes, the subject's brain behaved as if the person was having an actual experience.

The third row of the scan is where the deepest, most controversial, and perhaps most empowering implications of the mirror neurons for our lives becomes apparent. Here the subject is neither reaching out physically for the object nor watching someone else do any reaching. In this third part of the experiment, the subject is simply *visualizing* themselves in the act of reaching for the object. In their mind's eye, they are seeing themselves—imagining themselves—reaching for an object.

As we saw previously, while the degree of firing in the brain is somewhat reduced compared to the act of physically doing the reaching, clearly the brain is responding to the imagined experience as if it is engaged in the real-time, lived experience. This is the part of the experiment that opens the door to the more uncertain yet most potentially empowering of its implications.

The experience of using our imagination to visualize an action within our mind's eye evokes a similar response in our brain, and releases the same chemicals in our body, as if we are actually living the experience. It's for this reason that the way we see ourselves in the world—*the way we visualize ourselves both consciously and subconsciously*—is so significant for our health and well-being.

The body is producing the health, fitness, and vitality that we are programming it to produce through our self-image. And the perceptions of ourselves that we "feed" to our mirror neurons on a daily basis is a powerful part of the programming.

**PURE HUMAN TRUTH 39:** The way we see ourselves in our imagination triggers our mirror neurons to signal the body with the information to support our self-image.

The vital systems of the body, including the endocrine system and the immune system, respond adaptively to signals from the mirror neurons. They produce the chemistry in the body that matches and supports the conditions that we have observed (imagined) in our mind's eye.

With the Salk Institute experiment and similar experiments in mind, what would it mean for us to see ourselves as whole, complete, and healthy? And what are the implications of making this kind of visualization a practice on a consistent basis, day in and day out, holding that vision as part of a conscious ritual of vitality and healing?

The answer is clear. If the brain doesn't know the difference between lived, observed, and imagined experiences, then our ability to hold images of ourselves in mind as healthy beings, fully enabled and capacitated in every way, triggers our mirror neurons to respond in kind. And this is precisely what previous studies have illustrated.

In the past, the effect of visualization was attributed to what we know today as the *placebo effect*. We typically see this phenomenon as a mysterious, positive outcome of a health crisis that conventional medicine can attribute to nothing other than a patient's belief in their own capacity for healing, in a treatment, or in their caregiver.

With these discoveries in mind, my question to you becomes: What are you feeding your mirror neurons? What images of yourself are you creating, holding, and focusing on day after day, year after year? Do you focus on the things that you least like about yourself, such as the appearance of new wrinkles, a layer of body fat, a recent blemish, or do you hold an image of yourself in the optimal state of vitality and health that you desire?

While there are no "right" or "wrong" answers to these questions, your answers may help you to understand why your life and your body perform as they do. They may also give you the insights you need to make positive changes when you choose to do so.

## AWAKENING OUR SUPERPOWERS

In addition to the discovery of mirror neurons and their potential in our lives, another recent discovery is shaking the foundation of what we've been led to believe about our limits and our possibilities. It's the discovery of the neural network in the heart mentioned earlier in this chapter.

During my school years, I was taught that the brain is the master organ of the human body. It's the brain, I was told, that sends the signals and gives the instructions to the rest of the body that make it function in ways that keep us alive, healthy, and thriving. I was also taught that the heart is simply a pump—a really good pump—whose job it is to move blood through the body over the course of our lifetime. But new discoveries now tell us that both of these assumptions are wrong, or at least they are incomplete.

A growing body of scientific evidence now suggests that the pumping action of the heart, as important as this function is, may pale in comparison to the additional functions that the heart performs. In other words, while the heart does an astonishing job of pumping blood through the body over the course of a lifetime, the pumping may not be its primary purpose.

In 1991, a scientific discovery published in the journal *Neurocardiology* put to rest any lingering doubt that the human heart is limited to the functions of a pump. The name of the journal gives us a clue to the discovery of a powerful relationship between the heart and the brain that had been unrecognized previously. A team of scientists led by J. Andrew Armour, M.D., Ph.D., of the University of Montreal, that was studying the intimate relationship between heart and brain, discovered that about 40,000 specialized neurons, or *sensory neurites*, form a communication network within the heart itself.[15]

In simple terms, Armour and his team discovered the neural network that has now come to be known as the *little brain* in the heart. Their paper, published by the Royal College of Psychiatrists, reports in easy-to-understand terms: "The 'heart brain' is an intricate network of nerves, neurotransmitters, proteins, and support cells similar to those found in the brain proper."[16]

**PURE HUMAN TRUTH 40:** A neural network, or "little brain," within the human heart thinks, feels, experiences, and remembers independently of the cranial brain.

A key role of our heart's neural network is to monitor and detect changes in levels of hormones and other chemicals within our body, and then to communicate those changes to our brain. Our brain will then activate the systems that meet our needs accordingly. The heart fulfills this function by converting the emotional language of our body's perceptions, which includes the way we feel about our environment, into the electrical language of the nervous system so that its messages make sense to the brain. The heart's messages tell the brain when we feel safe in the world so we can focus on building a stronger immune system, for example; or when we feel that the world is not safe and need more adrenaline to adjust to a stressful situation.

Now that the little brain in the heart has been recognized by researchers, the deeper role it plays in performing physical and spiritual functions has also been revealed. A sampling of these functions includes:

- Communicating with sensory neurites that are located in other organs in the body.
- Accessing a form of deep intuition known as *heart intelligence*.
- Intentionally entering states of deep intuition on demand.
- Accessing the mechanisms of intentional self-healing.
- Triggering our abilities for super learning, super recall, and super cognition on demand.

## COHERENCE ON DEMAND

The little brain in the heart functions in two distinct yet related modes. In one mode, the heart can sometimes act independently from the cranial brain. When it does, it thinks, feels, learns, remembers, and even senses our inner and outer worlds on its own. It stores its impressions and memories locally, within the cardiac neural network.

In another mode, the heart can act in harmony with the cranial brain in a way that allows us to do something that no other form of life can do—at least the forms of life that we know of today. We can harmonize the neural networks in the heart and the brain to synchronize these two organs into a single, potent system that we can access and apply in our lives.

The three-step technique used to achieve the coherence of heart and brain is to:

1. Focus our awareness in the area of the heart.
2. Slow our breathing to a ratio of a longer exhaled breath than inhaled breath. For example, breathing in for a count of four, then releasing the breath for a count of six.
3. Consciously and intentionally evoke a positive feeling, such as gratitude or care, for anything or anyone.

The combination of these three steps, with the feeling created in Step 3 held for a minimum of three minutes, creates an electrical signal that travels from the heart to the brain and creates a state of harmony called *coherence*. When the feeling in the heart creates a signal that is 0.1Hz, this sets the stage for the optimal harmony that is simply called *heart-brain coherence*.[17]

Once again, I'm going to emphasize that we have the ability to self-regulate the inner conditions that create heart-brain coherence, because it is sometimes easy to gloss over the words that describe this process. We are the only form of life with the ability to consciously *choose* when we harmonize the neural networks of our heart and brain. We are also the only form of life that can consciously *choose* where, as well as how, we harmonize the power of two separate organs into one vital and powerful, networked system. The ability to establish this state of consciousness, at will and on demand, leads to a powerful level of mastery unique unto itself.

**PURE HUMAN TRUTH 41:** We are the only known form of life that can intentionally harmonize the neural networks of its heart and brain to create an optimized state of heart-brain coherence.

The act of creating heart-brain coherence optimizes the systems of our bodies, such as the immune system, the respiratory system, the nervous system, and others. In doing so, we become the best version of ourselves for whatever activities we choose to engage in throughout the day and to cultivate long-term well-being. The benefits of this synchronization are numerous and could easily be the topic of a whole book.

A 2022 article published by Alleviant Integrated Mental Health summarizes some of the benefits, stating: "As a result [of creating heart-brain coherence], you will feel more joy, increased energy and vitality, better decision-making abilities, and improved mental and physical health outcomes." (author's brackets)[18]

Some of the strongest evidence supporting the purely physiological power that heart-brain coherence affords us is found in the initial studies done by the HeartMath Institute, a research organization pioneering the study of the human heart in unconventional ways. HeartMath studies show, for example, that dehydroepiandrosterone (DHEA) levels in the body increase 100 percent over a 30-day period of establishing regular heart-brain coherence.)[19]

DHEA is a hormone that plays a vital role in our bodies. It's a precursor that must be present for our bodies to create and use other vital hormones, including testosterone and estrogen. Studies show that DHEA levels typically decline with age and must be available in the body to keep us healthy and strong. A regular practice of establishing coherence between the heart and the brain supports healthy levels of DHEA, even later in life when the levels typically decrease.

In addition to elevating DHEA levels in the body, the studies revealed that the levels of the stress hormone cortisol decreased 23 percent from the coherence created between the heart and the brain during the same time that the DHEA levels were increased. Further studies revealed that the antibody protein immunoglobulin A (IgA)—the first line of our immune defenses, which is located in the cells of the mouth—responded to a surprising degree for subjects participating in the research.

Following only five minutes of experiencing positive feelings that are part of the three-step coherence technique taught above, such as gratitude and care, the subjects experienced an "immediate 41 percent average increase in their IgA levels. After one hour, IgA levels returned to normal, but slowly increased over the next six hours."[20]

## RECOGNIZING OUR SUPERPOWERS

The significance of studies such as those conducted by HeartMath and Allegiant is twofold. First, they tell us that we have the ability—the power—to awaken and enhance important physical responses, such as our immune response, deliberately. Systems positively affected by the state of coherence are systems that were previously thought to be beyond our ability to access in a meaningful way.

Second, the fact that it's possible to enhance our bodies as the result of experiences that we can manage and create when we choose to create them—on demand—tells us that we are imbued with the ability to self-regulate our extraordinary soft technology, including all of the functions described previously in this chapter. These are the kinds of abilities that engineers dream of imbuing into artificial systems, such as robots and AI applications.

Armour's discovery of neurons in the heart gives us reason to change the way we think of ourselves forever. It gives new meaning to what's possible in our bodies as well as to what we're capable of achieving in our lives. The science from the new field of neurocardiology is just beginning to catch up with ancient, Indigenous, and spiritual beliefs when it comes to explaining the source of experiences such as intuition, precognition, and self-healing. Almost universally, traditional teaching acknowledges the heart's direct influence upon our personalities, our daily decisions, and our ability to make moral choices that include the discernment of right and wrong.

Embracing the benefits of the heart's wisdom can immediately catapult us beyond the limits we've perceived for ourselves in the past when it comes to the way we live, our capacity to solve problems, and perhaps most significantly, our capacity to love. It's these capabilities as well that give us the resilience to embrace big change in our lives, and to do so in a healthy way.

When we take into consideration all that we now know about the heart, including the fact that it is part of an extended neural network that had already developed when our ancestors first appeared on Earth, the fact that we have a little brain in our hearts made of cells that think, feel, and remember independently of the cranial brain, and the fact that we can self-activate the benefits that come from the relationship between the brain and heart, then the questions we must ask ourselves are: What else does the heart do that we're only beginning to understand? What capabilities await discovery today that we've either forgotten we possess or are only just now beginning to understand fully?

Lately, new discoveries published in scientific journals are answering these questions on almost a weekly basis.

## AWAKENING OUR SOFT TECHNOLOGY

Discovering the relationships between the cells of our body, mirror neurons in the brain, and the power of the human heart has paved the way to our understanding of a new kind of technology beyond anything we've conceived of in the past. The technology is us. We are the sophisticated soft technology that we strive to build into the most advanced machines of our day. To see ourselves this way requires a kind of thinking that is very different from viewing the human body as a fragile system of lucky biology, saline fluids, blood, and bones.

In a very real sense, our soft technology is light-years beyond anything modern engineers are capable of building in laboratories and robots today. Soft technology means that we're beyond the primitive components of hard silicon computer chips, embedded nanobots, and sensors implanted into the skin.

We're more than the artificial proxies for flesh-and-blood organs, which have only a limited potential to mimic our natural abilities. We're more because we're a different kind of technology. We're a soft technology. We're made from neurons that have self-healing membranes and ion potentials moving across cell walls that are turned on, shut off, and self-regulated, at will and on demand when we choose to awaken these systems. This is what it means to be pure human.

Rather than viewing the human body as the flawed product of random mutations occurring over a long period of time (as Charles Darwin's theory of evolution asks us to accept that it is), in fact, it is an advanced form of biological technology. If we allow ourselves to see our humanness through the lens of engineering, we begin to realize just how sophisticated and intricately designed our biological and cellular structures and potentials are. Everything we see engineers working to invent and further develop in the world around us, from advanced communication networks to blockchain technology, is a poor approximation of the communication networks and growth, diagnostic, and repair systems that have always existed within us.

The key here is that we already do the things and perform the functions that the transhumanistic proposals ask us to accept into our bodies, only we do them better.

**PURE HUMAN TRUTH 42:** The pure human body is a highly advanced, technologically sophisticated soft technology that, in many ways, outperforms artificial components of computer technology to make possible advanced states of consciousness and healing.

We can think of the organs and systems in our bodies as separate but harmonious programs running under the operating system of consciousness, which gives us access to our divinity. We can also think of our divinity as a guidance system that steers us to become the best version of ourselves. When we consider our humanness from these perspectives, in a very real sense, we are the advanced technology that the best engineers of our day dream of creating. We are the technology we've been waiting for.

CHAPTER FIVE

# Everyday Divinity

## Our Destiny of Living Free

Each person is born with an infinite power against which no earthly force is of the slightest significance.

— Neville Goddard (1905–1972),
Bajan writer, speaker, and mystic

The key to awaken the extraordinary power of our divinity is beyond the words of the stories that we tell ourselves and believe. We can say to ourselves, for example, that we are born with super abilities of deep telepathy, self-healing, and longevity that rival the power of the gods. We can attend seminars and workshops, register for online courses, and travel to remote and mysterious places in the world that preserve the legacy of past civilizations in our search to discover our own superpowers. While each of these actions may inspire us and lead to a community of like minds that supports us in our journey, ultimately they may fall short of fulfilling our dreams.

Until we actually *live* the powers that we discover and take responsibility for implementing them in our lives, they remain little more than wishful possibilities that we aspire to. Just as the output of a computer program is more than the language used to write it, we must search deeper than the words of our stories to know the potential and healing that awaits our embrace.

It may be precisely this power that the 19th- and 20th-century Armenian-French philosopher and mystic George Ivanovich Gurdjieff, commonly known as Gurdjieff, discovered as the result of his search for deeper meaning in his life. In his memoir, *Meetings with Remarkable Men*, Gurdjieff details his years of exploration to uncover the mysteries of our past. His journeys led him throughout Egypt, Central Asia, India, Iran, and Tibet as he followed ancient clues that led him from temple to village, and teacher to teacher. Eventually he found himself in a secret monastery hidden in the Hindu Kush mountains of northern Afghanistan, now believed to be the home of the mysterious Sarmoung Brotherhood.

It was there that the abbot of the monastery offered words of encouragement to him that made his search worthwhile. In the film adaptation of Gurdjieff's story, the actor portraying Gurdjieff says, "You have now found the conditions in which the desire of your heart can become the reality of your being. Stay here until you acquire a force in you that nothing can destroy."[1] I have no doubt that the power of human divinity—the ability to transcend our own perceived limitations—played a powerful role in the conditions that Gurdjieff had discovered.

To unleash the conditions in which our hearts' desires become the reality of our lives, we must understand our relationship to ourselves, our world, and ultimately to God. Through the words of our past, we are given the knowledge of how to do just that. In his book *The Prophet*, Lebanese-American poet Kahlil Gibran reminds us that we can't be taught the truths we already know. "No man can reveal to you," he states, "that which already lies half asleep in the dawning of your knowledge."[2] It makes tremendous sense that hidden within us we would already have the knowledge and the ability to awaken the force that allows us to "transcend our perceived limitations"—the very definition of achieving our divinity. It's through the greatest expressions of our divinity that we achieve the highest levels of our life's destiny.

**PURE HUMAN TRUTH 43:** The greater our expression of divinity, the more of our destiny we achieve.

While we may draw upon a variety of sources to provide inspiration and direction in life, including religious and spiritual guidance, those sources cannot do the work for us. To achieve the realized destiny of our potential fulfilled is an inside job. It is only we, through our life choices, who can allow ourselves to be inspired and act upon the guidance we receive.

When it comes to the basic structure and biological "wiring" that gives us our humanness, we are all pretty much the same. The basic components of an advanced neural network with a capacity for elevated brain states, such as mirror neurons, and the self-regulation capacity of our cellular biology, are already present. And while our unique life experiences may have resulted in some of us having portions of our wiring, and thus our abilities, more developed than others, barring extreme circumstances that may incapacitate us in some way, each of us has the capacity to uplevel, fine-tune, and hone our skills of divinity.

The purpose of sharing the following examples is to inspire, awaken, and affirm this realization in you, and in myself, using concrete examples of lived human experiences. By admiring other people's accomplishments, we open the door to the possibility that what we've often thought of as rare and supernatural experiences happening only to special people in the past may in fact be showing us what's possible and normal for all of us as a natural part of our everyday lives.

## A MOTHER'S INTUITION

In the spring of 2012, T.J. Findlay, a private in the Australian Army, was on a routine patrol with his military unit in a remote province of Afghanistan. Suddenly, he was blown off his feet from the shock wave of a violent explosion that occurred on the road in front of him. Stunned, but still conscious, he looked up from the ground where he'd been thrown and saw the vehicle in front of him in flames. His convoy had either been the target of a roadside bomb or possibly taken the direct hit of a rocket-propelled grenade. A passenger in the burning vehicle in front of him was an Afghani soldier, a friend with whom T.J. had recently bonded, as they were both fathers of young children that they missed during their deployment.

Instinctively, T.J. ran toward the burning vehicle to pull his friend and any survivors from the fire as quickly as possible. Due to a combination of T.J.'s training and his heroism, both he and his friend miraculously survived the attack in relatively good condition. Later that day, on a videoconference call with his family in Brisbane, T.J. recounted the harrowing experience to his wife, Kira. It was during his call that he learned that both his wife and his mother already knew that he had been in danger.[3] And while they didn't know the specifics of what had happened, they didn't have to, to know that something was wrong.

Earlier that day, T.J.'s mother had experienced a strong feeling that something had happened to him. Without having the details of her son's brush with disaster, she knew with absolute certainty that something was terribly wrong. Through the intuitive connection that she had known and felt with T.J. from the time of his birth, she knew that he was in trouble. Relying on the power of their shared faith, his mother and Kira had both been in prayer for T.J.'s health and safety throughout the day. It was only after he contacted them later in the day that their intuition was validated and they learned the details of T.J.'s close call.

Fortunately, this story of a mother's intuition regarding her son during a time of war had a good outcome. The question is this: How did T.J.'s mom know with such certainty that her son was in trouble? How could she have detected that he was in danger before receiving his call to the family or any notification from his commanding officers?

## CONNECTED FOR LIFE

We often hear stories of a mother's connection with her children and of how that connection can continue beyond their birth and childhood even into their adult years. Mothers will commonly say that they "know" instinctively when their baby is hungry, when they are hurt, and when they need help. Until recently, the relationship between a mother and her offspring, both child and adult, has been largely anecdotal. New discoveries have changed that and now give us insights into how maternal intuition works, how it can be strengthened, and why it lasts until so much later in life.

A 2015 paper published in the journal *BioEssays* details a study regarding the relationship between a mother's body and the DNA of the fetus in her womb, with a surprising outcome.[4] The research revealed that the DNA of an unborn baby is not confined to the body of the fetus itself. Rather, while it is in the womb, the baby's DNA will circulate in the blood of the mother's body. And the key insight here is that the DNA remains in certain organs and tissues of the mother's body for many years after pregnancy and delivery. Specifically, the paper states: "Fetal cells have been found to stay in the mother's body beyond the time of pregnancy, and in some cases for as long as decades after the birth of the baby."[5]

An unexpected outcome of the study, and one that may help us to understand the mechanisms behind the power of intuition, is that the sharing of the baby's DNA with the mother is a two-way experience. While the DNA from within the womb is circulating through the body of the mother, the mother's DNA is also circulating and lodging itself into the organs and tissues of the unborn baby at the same time.

Specifically, the article reads: "The mom's cells also stay in the baby's blood and tissues for decades, including in organs like the pancreas, heart, and skin."[6] It is this two-way exchange that helps us to understand the intuitive connection between a mother and her child.

## OUR DNA ANTENNA

Earlier in the book, we discovered that, in addition to providing the blueprint for the development and growth of our bodies, DNA also functions as an antenna that receives information from the intelligent field of energy/information that underlies our existence. Specifically, the paper stated that "3D-A-DNA structure behaves as a fractal antenna, which can interact with the electromagnetic fields over a wide range of frequencies."[7]

It's the antenna function of DNA that gives us insight into the deep intuition that exists between a mother and the child. Both when the baby is inside of her body, and lasting for a period of time now measured in decades following birth, mother and child are communicating via nature's "Internet" for sharing information through biological resonance.

It also appears to be this connection that contributes to the pain we feel when we as adults lose our parents, or when a parent tragically loses a child. We say that we feel a sense of loss and emptiness, and often even a physical pain in our heart and gut that conventional medicine does not have the tools to explain.

On a personal note, when I lost my mom to Covid-19 during the pandemic in 2021, I was hospitalized for cardiac symptoms that appeared to have no biological source. All of the physical tests and chemical markers came back showing perfect readings. The mysterious symptoms were ultimately attributed to the somatic conditions produced from the grief—the unresolved grief—I felt over the loss of my mother. With this idea in mind, perhaps it's not surprising that the physical symptoms immediately disappeared when I began to directly address my grief specifically. To this day, those symptoms have never returned.

The point of me sharing this account is that we now have a biological and physics-based understanding of the maternal intuition that used to be anecdotal and attributed to coincidence. As we accept the fact of our energetic connection, it becomes easier to acknowledge that this connection is not limited exclusively to mothers and their children. The reason is that every strand of DNA throughout the 23 pairs of chromosomes located in each nucleus of the approximately 50 trillion cells that make up the average human body are antennae. And they are tuned to something. While some of the antennae remain "tuned" to the mothers that carried us, other portions of the DNA antennae are tuned to elements of the world we live in.

## WHAT ARE WE SAYING TO THE FIELD?

Through our resonant relationship with the energy field that underlies our existence, we are in a never-ending conversation with the world around us. We are constantly sending vibratory messages into this field that is the source of our experience. At the same time, we are constantly receiving information and being affected by the same field. When we really acknowledge the existence of this two-way communication dynamic, and the role that the field plays in our lives, the obvious question becomes, what are we saying to the field? To develop an awareness of our thoughts, emotions, and beliefs and the way that we respond to the events of our lives is to become conscious of the language that we are conveying into the source of our existence. Awareness and mindful noticing are also factors in becoming aware of what the field is saying to us through our feelings and intuition.

**PURE HUMAN TRUTH 44:** We are in constant, resonant communication with the world around us through the biological antennae of our molecules and cells.

We don't need to be mothers to receive the intuitive "hits" that are coming at us each and every day. Such information is a natural and inescapable part of our lives as a consequence of living in the field. Some of our highest levels of mastery may be gained through embracing these expressions of divinity and allowing intuition insights to become a natural part of our everyday lives.

## DOING THE "IMPOSSIBLE"

In 1863, British runner William Lang was documented to have run a mile in four minutes and two seconds. For 91 years following Lang's record, all attempts to run a mile in under four minutes, although they often came close to that speed, were unsuccessful. Using different surfaces for the running tracks, varied training methods, and different kinds of shoes, for nearly a century, it appeared that the human capacity to run a mile in under four minutes represented some kind of a mysterious limit for our muscles, coordination, and speed. In the spring of 1954, all of that changed.

On May 6 that year, another British athlete, 25-year-old Roger Bannister, ran faster than anyone had been documented to do in nearly a century's worth of track-and-field events. In doing so, he shattered the seemingly impossible record for the four-minute-mile. His time was three minutes and 59.4 seconds. As amazing as Bannister's achievement was, however, something even more amazing happened soon after he set this world record. The events that followed Bannister's record-breaking time are the reason that I'm writing this chapter.

Only 46 days after Bannister set his record for the mile, Australian runner John Landy broke the record again, running the distance even faster than Bannister had done and breaking Bannister's record by an even greater margin. Landy's time was a sizzling three minutes, 57.9 seconds. In the world of athletics, what may appear to some people as simply a few fractions of a second for Landy represented a stunning chasm between his time and Roger Bannister's. But the record-breaking results did not end with Landy. In the decades that have passed since his achievement, as of 2022 another 1,755 athletes from around the world have now broken records for running the four-minute mile. As of July 2024, the record is 3:43.13, and the tally for those who've beaten the clock includes 109 women.

## SOMEONE MUST DO IT FIRST

The sequence of events that led to the now-common running time of a sub-four-minute mile gives us a powerful insight into the world, and our lives, as well. After Lang's record was set, it took nearly 100 years for the first human to run a mile in under four minutes. Once it happened, however, it took only 46 days for someone else to break the new record. Today, the ability to run a mile in less than four minutes is no longer seen as an impenetrable barrier to human capacity.

Through new forms of physical, emotional, and psychological training, runners now routinely achieve sub-four-minute miles. And, interestingly, the times are getting faster and faster. The question is, why? Why wasn't the four-minute mile record broken earlier, in 1900? Or in 1926 during the running of the prestigious Wanamaker Mile in New York City? Or during the United Kingdom's Emsley Carr mile that began in 1953? And why has it become easier for athletes to do so now, after Roger Bannister broke the four-minute barrier in 1954?

The answer to these questions is brief and simple. The bottom line is that *someone had to do it first*. One individual had to accomplish the seemingly impossible feat first, and do so visibly and publicly in the presence of people, before others could follow. And this is the point of sharing this account. For reasons that we'll explore in this chapter, we typically must first witness the extraordinary act, feat, or accomplishment that we aspire to do already fulfilled, before we can accept that it is possible for us to do it in our own lives.

For example, if Roger Bannister had broken the four-minute barrier secretly, perhaps on an indoor track tucked away in a part of the world that no one knew about, or he did so in a room with no windows, no cameras around to record the feat, with no one present to witness his accomplishment, it would not have had the same impact on other runners.

While he might still have broken the record, if there was no one to know about it or see him doing it, the unseen act would not have had the same ability to impact the perceptions and expectations of other runners. This is the key to expanding what we view as limits to our capabilities, and to the *destiny code* to which each of us is subject.

## THE POWER OF WITNESSING THE IMPOSSIBLE

When we witness someone who is essentially like us accomplish an extraordinary feat—another human who is born as we are, lives and breathes as we do, eats food like we do, and has feelings and emotions just as we do, yet this person does something that we've believed was unattainable in the past, then this person's exceptional accomplishment becomes a bridge that links us to a greater possibility in our lives.

On a nonverbal level, the act of witnessing someone else's accomplishment seems to give us a form of psychological permission, saying to us that if another can do the feat, then maybe we can do it as well. This is not the kind of permission that comes from an authority figure who has power over us. Rather, it is an emotional and psychological permission code that speaks to us, breaking through our own barriers of disbelief and self-doubt. Knowing that such bridges exist is the key to fulfilling our human destiny.

**PURE HUMAN TRUTH 45:** When we witness one person succeed where others have failed, their accomplishment is a bridge that makes it easier for us to follow in their footsteps.

As we discovered in Chapter Four, we are born with the extraordinary potential to move the electricity in our cells, and to store and concentrate energy, information, and light in precise ways in our bodies. Our potential to do these things can sometimes be expressed as physical feats, such as running a sub-four-minute mile. It can also be expressed by achieving a seemingly miraculous healing within us ourselves, or by facilitating such a miraculous healing for another person.

Sometimes we experience our potential as the ability to perform seemingly superhuman feats of telepathy, psychokinesis (moving physical objects without touching them), super learning, super memory, and super cognition. And while some people appear to be born with the tendencies to accomplish extraordinary feats easily, others can often learn the same skills by being in the presence of those who have already mastered them.

Ancient yogic masters in the Himalayan mountains, for example, demonstrated to their students that they could transcend the "laws" of physics using nothing more than breath and emotional focus. Once the apprentices witnessed their masters' supernatural powers of healing or bilocation, they had to make a choice as to what the experiences they'd seen meant in their lives. They had either to discount what they'd seen altogether and chalk them up to illusion or magic, or to accept what they'd seen and adjust their lives and belief systems to make room for a new reality.

The following story of the great yogi Milarepa, who lived circa 1100 C.E., and my personal experience of the legacy he left for his students, is a perfect example of what I mean here.

## FREE FROM PHYSICAL MATTER

In the 9th century, the Tibetan Buddhist yogi Milarepa began a personal retreat in the isolated caves of the Himalaya mountain range. The purpose of his seclusion was to redeem himself following his abuse of the martial arts powers that he had mastered earlier in his life and the suffering that his abuse of these powers had brought to the lives of others. He felt that by dedicating the rest of his life to fully mastering the untapped potential within his body, and by using that potential only to do good things in the world, he would honor his purpose in life and earn forgiveness for his misdeeds.

His journey of redemption lasted until his life ended at the age of 84. Without any substantial food, clothing, or distractions to interrupt his inner focus, for years Milarepa existed eating almost nothing, with almost no possessions, in the harsh cold of the high-altitude environment. His only human contacts were the students and occasional pilgrims who would stumble upon the caves that sheltered him. The deprivation that he imposed upon himself, although extreme, ultimately facilitated his ability to attain his life goal of yogic mastery.

Before his death in 1135 C.E., Milarepa left proof of the freedom that he achieved from the physical world. That proof remains today in the form of a series of miracles that continue to baffle modern scientists. The existence of these miracles represents a phenomenon that scientists say is impossible and cannot exist. And yet the miracles remain. The fact that they do is the reason that I'm sharing this account.

To demonstrate his yogic mastery over the confines of his physical environment—the cold rock surfaces of the cave walls that surrounded him—Milarepa repeatedly performed a feat that scientists and skeptics have never been able to duplicate. He would begin by placing his open hands against one of the rock walls at shoulder level. He would then continue to effortlessly *move his hands further into the rock that was located either in front of him, above him, or below him, as if the rock didn't exist*! When he did this, his students reported that the cave walls beneath his palms behaved as if they were melting under his touch.

As Milarepa continued to push into the cave's walls, floors, and ceilings, he left the deep impressions of his hands, and his feet as well, for all who came to see. Those impressions remain in Milarepa's caves today, and continue to inspire present-day yogis.

## SEEING FOR MYSELF

In 1998, I led a multiweek group pilgrimage onto the Tibetan Plateau and intentionally chose a route through Nepal that would lead to Milarepa's caves. As both a scientist and a martial artist, I had studied the yogi's life and traditions for years and I wanted to see for myself the legacy of his mastery and the evidence he'd left for future generations.

After 19 days of travel and acclimating to elevations of 13,000 feet above sea level, I found myself standing in the great yogi's retreat, precisely where he had stood nearly 900 years before. With my face only a few inches away from the wall of the cave, I was staring squarely into the mysterious hand prints that Milarepa left behind.

I opened the palm of my right hand and placed it into one of the impressions of Milarepa's right hand. I could feel my fingertips cradled in the imprint left behind by the yogi, in the precise position that he'd placed his fingers eight centuries before—a feeling that was both humbling and inspiring at the same time. The fit was so perfect that any doubt I had about the authenticity of the handprint quickly disappeared.

Immediately, my thoughts turned to the man himself. I wanted to know what was happening to him when he merged with that rock. What was he thinking? *What was he feeling?* How did he defy the physical "laws" telling us that two "things"—like a human hand and a rock wall—can't occupy the same space, at the same time?

The Tibetan translator who was with me during this visit seemed to anticipate my questions and volunteered that Milarepa strongly believed that he and the rock walls of the cave were not separate. From his yogic practices, he learned that the rock could not contain him or limit his movements. Milarepa demonstrated his belief to himself and to his students, by moving his hands freely wherever he chose, even if the wall of a cave happened to be in the place of his movement.

## BEING CHANGED IN THE PRESENCE OF THE IMPOSSIBLE

When Milarepa's students saw him do something that conventional beliefs said could not happen, it inspired them. His feat gave them permission to break free of their limited beliefs regarding what was possible for them to do.

This is the very definition of a divine power: it enables us to *transcend* our perceived limitations. The students witnessed the teacher's mastery of solid matter with their own eyes. In this case, it wasn't through the words of a legend passed from one generation to the next that they discovered the reason to shift the way they thought of their relationship with the world. They witnessed the reason for themselves. They experienced the feat directly and personally. And because they personally saw the miracle occurring, it was their experience that told their minds they were not bound by the laws of reality as they were known at the time.

**PURE HUMAN TRUTH 46:** When we witness another person transcend limitations that we've accepted for ourselves, we must choose either to discount what we've seen, viewing it as unattainable, or to accept what we've seen and shift our belief system to accommodate the new belief.

Although we live in a time 900 years after Milarepa, our lives today are not so very different from the students' in the cave. We each face the same dilemma that the students faced when they saw the yogi's accomplishments. The dilemma is this: On one hand, the families, friends, and loved ones of Milarepa's students lived in a time and place where they accepted a certain way of seeing things and how the world works. This included the belief that the rock of a cave wall is a barrier to the flesh of the human body. On the other hand, the students had witnessed an exception to the "laws."

The irony is that both ways of seeing the world were absolutely correct. Each one depended upon the way that someone chooses to think of their relationship to the world, in a given moment of time. It was about what they accepted and believed.

Using the language of quantum physics rather than yogic miracles, a growing number of leading-edge scientists suggest that the universe, and everything housed in the universe, is what it "is" *because* of the force of consciousness itself: our beliefs and what we accept as the reality of our world. Interestingly, the more we understand the relationship between our inner experiences and our world, the less far-fetched this suggestion becomes.

While the story of Milarepa's handprints is a powerful example of one man's journey to discover his relationship to the world, we don't need to make a pilgrimage to the Himalayan mountain range to discover the same truth for ourselves! Modern-day examples of miraculous events that defy our current scientific paradigm and challenge our belief systems are available to us today.

And just as the students of masters past had to come to terms with the power of those miracles and what they meant in their lives way back when, we have the opportunities to do so today as well. The documented healing of a seemingly incurable condition using an advanced form of energy work has opened the door to new possibilities for the thousands of people who have traveled to Milarepa's caves or seen the evidence recorded on video.

## MIRACLE OR TECHNOLOGY?

In 1995, I had the opportunity to witness a modern-day miracle few people in the Western world had seen at the time. I was studying Chi-Lel, a form of qigong that had just been introduced to the Western world through the lineage of students who had carried this form forward in China for 500 years. The miracle was the disappearance of a life-threatening tumor inside the bladder of an adult woman in a way that was immediate, verifiable, and occurred in real time. Western doctors had diagnosed the mass in the woman as malignant and told her that it was inoperable. She thanked them for their opinion, and began the journey of alternative healing protocols that ultimately led her to the experience that I was about to witness.

During a course that I was attending with a small number of dedicated students, our group was shown a film that was created by our instructor. He had been asked to record the procedure when he was present for the healing that happened in what is known as a "medicineless hospital."[8] This particular hospital was located in Beijing, China.

The Beijing clinic was one of many in the region at that time that routinely used traditional, nonmedical methods of healing, and did so with predictable success. After the instructor created a context so the film would make sense to us, we were prepared for what we were about to see. The instructor emphasized that the purpose of the film was to show us that the power to heal is something that already lives within each of us. It was *not* an advertisement for the clinic, or an invitation for everyone with a life-threatening condition to make a mad dash to Beijing.

What we were about to see in the film could be accomplished right there in our classroom, or at home in the living rooms of our families and friends. The key to healing, he said, is the ability to focus emotion and energy in our bodies, or in the body of a loved one (with permission), in a noninvasive and compassionate way.

The woman in the film had come to the medicineless clinic as a last resort, because all else had failed. The clinic first emphasized personal responsibility for the woman's everyday health. She was required to participate in new and life-affirming ways of living, moving, and breathing, rather than having the staff at the clinic simply "fix" her and send her home. These preliminary steps were vital for the healing. Without them, the healing that she was about to experience could be only temporary, or possibly ineffective. The life-affirming protocols the woman learned included new ways of eating, gentle forms of movement to stimulate the life force (*chi*) within her body, and special ways of breathing that promote physical healing.

By following these simple changes in lifestyle for a few weeks before the actual procedures done at the clinic, the woman's body was strengthened for the natural healing that was possible. Following these procedures, it then made sense for the woman to undergo the treatment that was recorded on the video.

As the handheld film began, we could see the woman with the tumor lying on what appeared to be a hospital gurney. She was wide awake, fully conscious, and had been given no sedatives or anesthetics. Three practitioners in white lab coats stood behind her, while an ultrasound technician was seated in front of her, holding an ultrasound wand that would be used to create a real-time view of the mass inside her body. We were told that the image would not be time lapsed, like an educational nature film showing, for example, the days-long process of a neural network forming, condensed into only seconds. Our film would be unedited so we could see the true effect that the practitioners had upon the healing in real time.

The film was short, lasting less than four minutes. Within that time span, we all saw something that is considered a miracle by modern science and the standards of Western medicine. Yet, within the context of the information I've shared in this book, it is something that makes perfect sense. The practitioners had agreed upon a phrase that would reinforce a special kind of feeling inside of them. When I learned the English translation of their phrase, it reminded me of a principle revealed by the 20th-century Bajan philosopher Neville Goddard when he said, "Make your future dream a present fact . . . by assuming the feeling of your wish fulfilled."[9] The practitioners' feeling was the powerful, focused, and felt sense that the woman was already healed.

Although they knew that the tumor had physically existed in the moments leading up to the process, the healing practitioners also acknowledged that its existence was only one possibility out of the many that exist in the quantum field. On that particular day, they awakened the code that calls into existence another possibility. And they did it in the language that the field recognizes and responds to: the language of human emotion directing energy.

Watching the practitioners, we heard them repeat the words of a mantra of sorts, which loosely translates into the English words *already done, already done*. Our little classroom was absolutely silent as we watched the events that unfolded on-screen with awe. At first, it seemed that nothing was happening. Suddenly, in real time, the tumor began to quiver in and out of view, as if it were teetering between realities. Within seconds, it faded and then completely vanished from the screen. It was gone.

Everything else was still there, just as it had been only seconds before. Everything, that is, except the tumor that had threatened the woman's life. The walls of the woman's bladder were the same. The registration marks on the sonogram were the same. The room looked the same. The practitioners and the technician were there, and nothing "spooky" appeared to have happened anywhere else. That is, with the exception of the condition that had threatened the woman's life only seconds before.

The tumor no longer existed.

## MOVING MOUNTAINS

I remember thinking to myself the ancient spiritual admonition that with just a little focused faith—the faith of a mustard seed—it is possible to move mountains. I also remember thinking that before that moment, I had always believed that the moving of mountains was a metaphor. Now I knew that it was literal. The three Chinese healing practitioners had moved the biological mountain of the tumor and had done so without physically touching the woman's body.

Back in the 1970s, researchers studying the effects of Transcendental Meditation (TM) documented that real-world conditions can be influenced when a relatively small number of advanced meditators perform their meditations simultaneously.

Specifically, the researchers discovered that when as few as the square root of one percent of a given population meditated using the TM techniques, the population being studied experienced a decrease in crimes against individuals, fewer traffic accidents, and fewer hospital emergency room visits.

The formula revealed by the TM researchers for entire cities appeared to have been validated with the small population of the clinic on the day of the filming. There was a total of six people in the clinic video while the healing was happening (three practitioners, the technician, the camera operator, and the woman undergoing the healing.) Applying the formula, the square root of one percent of that room's population on that day was only .244 of a person! With less than one person's absolute acceptance and belief that her healing had already happened, the physical reality of the woman's body changed.

Though the numbers in this instance were small, the formula still held true. As noted before, the square root of one percent of a given population is the *minimum* number of people needed to jumpstart a new reality. In all probability, all six of the people in the room felt the feeling of the healing, and the video showed that it took two minutes and 40 seconds for her body to mirror their collective reality.

With these statistics in mind, after viewing the film in the classroom, I took the opportunity to ask our instructor if it was necessary for multiple people to be present for a successful outcome. Was it really necessary for three practitioners to be present? Would it have worked if there were only two? Or even one? The answer that the instructor shared with me offers a powerful conclusion to this story. "In all probability," he said, "the woman could have accomplished the healing even if no one was with her. *She could have healed herself.* However, because it's sometimes hard for us to accept our own power, we seem to do better when we are surrounded by other people who think and believe as we do." And there it was. The confirmation of what we all sense yet are often reluctant to claim in our own lives: we are immensely powerful beings with godlike abilities when we are tapped into the field of possibilities that underlies our existence.

The conditioning that we are weak, powerless, and largely unworthy of goodness and healing in our lives is so deeply etched into our psyches as the result of programming we get from family, society, and religious institutions that it is often difficult for us to accept our power even when it comes to our own healing. For this reason, Chi-Lel practitioners typically work with others to support their patients, as well as one another, during their healing sessions.

With permission from the practitioners, I have since shared this film with many audiences throughout the world. Without permission, there are times that audience members have recorded and shared excerpts of the video on social media, and done so irresponsibly, without explaining the context of what it is that's being shown. As you may imagine, this irresponsible practice has led to confusion for some and mistrust for others, believing they're seeing a "deep fake," a Photoshopped or AI-altered image. I'm sharing the meaning of the film here to remedy possible confusion about what this practice represents and what it means.

## WHY WOULD OUR HEALING BE ANYTHING LESS THAN EASY?

The reactions to the film from audience members vary, and are predictable. Once the healing happens, there is generally a brief silence as the audience registers in their hearts and minds what they've just seen with their eyes. The silence gives way to sighs of joy, laughter, and even applause. For some people, seeing the film is overwhelming. For others, it is a welcome validation of what they already sense is possible. Even their believing, however, is bolstered by the tangible validation of witnessing what is considered by many to be impossible, and seeing it occur in real time.

For other people who are more skeptical, a question I typically hear is, "If this is real, why don't we know about it?"

My answer is, "Now you do!"

"How long does the healing effect last?" is the next question. When I asked our instructors the same question, I learned that the follow-up studies show a 95 percent success rate after five years for clients who continue the life-affirming breath and movement practices and lifestyle modifications, such as nutrition, that they learned at the clinic.

With a gasp that comes from somewhere between wanting to believe and the frustration of knowing so many people who could not be helped by conventional Western medical techniques, I usually hear something like, "This is too simple . . . it just can't be that simple!"

But why would we expect anything less? In the field, all things are possible, and we choose our possibilities. Knowing that everything from the most horrible suffering to the most joyous ecstasy, and every possibility between those extremes, already exists as a potential, it makes perfect sense that we have the power to collapse the space between the extremes and bring a possibility of choice into our lives. And we have the power to do so through the silent language of our divinity communicating with the field through the vehicles of our imagination, dreams, and beliefs.

Why should our healing be anything less than simple and easy?

Believing that we are "here" and the possibilities are "out there" sometimes gives us the feeling that positive possibilities are inaccessible. The same rules that describe *how* the field works, however, also tell us that in a deeper reality, what we typically think of as "somewhere else" is really already "here," and vice versa. It's all about the way we think of ourselves in the field of possibilities.

The scientific discoveries of the last 150 years, such as the mysterious result of the double-slit experiment, showing how invisible waves of potential energy collapse into the visible particles of reality when we observe them, have already shown that consciousness, reality, and belief are intimately related. They also have shown that our brains and bodies are wired with nerve cells that enable us to make use of, and participate in, that relationship.

The ability to use consciousness to navigate the field that connects all things is a perfect example of a modern-day application of this ancient relationship that spans consciousness and our everyday world.

## WHEN "THERE" IS "HERE"

During the 1990s, I had the opportunity to tour as a keynote speaker with an international conference featuring some of the leading voices in breakthrough science. From former astronaut and sixth human to walk on the moon Edgar Mitchell, linguist and ancient Sumerian scholar Zecharia Sitchin, and physicist Michio Kaku, the co-developer of string theory, to UFO-abductee psychologist John Mack, I had the opportunity to speak directly to the thinkers who were leading the way in unveiling a new human story.

While each of the scientists and researchers brought a unique perspective to the emerging story of the universe, one of the most impactful when it comes to unlocking the mystery of our divinity was American physicist and parapsychologist Russell Targ. The topic that Targ was addressing during the tours was related to the phenomenon known today as *remote viewing.*

In the 1970s, Targ and his team at the Stanford Research Institute (SRI) at Stanford University were given the task of first validating the remote-viewing phenomenon and then developing a structured protocol that could be repeated, taught to others, and used in applications ranging from military projects to identify the location of hidden enemy weapons to law enforcement applications that included locating missing persons.[10]

A simple definition of remote viewing is the ability of an individual to achieve a state of consciousness that allows their awareness to observe distant objects, locations, and events in real time although the person doing the viewing never leaves their fixed position. During a remote-viewing session, the experiencer, or viewer, works with a trained facilitator who helps them drop into this specific state of consciousness and, once there, to remain in that state rather than slipping deeper into a more familiar sleep and dream state. Once the viewer has attained this level of consciousness, it becomes possible to maneuver their awareness and navigate to any location in the field to view the details of that place. At the level of existence achieved as this state of consciousness, the concepts of "here" and "there" break down. To the person navigating the field, here is there, and there is already here.

The target locations for a remote viewer can vary from familiar coordinates, such as latitude and longitude, to a prominent landmark, a physical street address, or even an office or apartment number in a building. As mentioned previously, through our biology we are linked to a field of energy and information that is not confined by the laws of physics, time, and space.[11] The relationship between our biology and the field means that the awareness of the remote viewer is not limited only to targets in the present day, or even on Earth. The experiments have demonstrated that it is possible to remote view events occurring in other parts of the cosmos as well as those that will happen in the future or have already occurred in the past.

## DEEP INTUITION CAN BE LEARNED

In 1972, the U.S. military officially entered into the arena of investigating deep intuition and remote viewing via the Cold War strategy of *psychic warfare*. The Central Intelligence Agency (CIA) discovered that the former Soviet Union had been experimenting with human telepathy and other psychic abilities for years, placing the United States at a disadvantage as a latecomer in the exploration. It was during this time that Targ and Harold Puthoff, his colleague at the SRI, were contracted by the U.S. Army to investigate what were called *repeatable psychic phenomena* and their possible military applications.

The initial investigations at the SRI led to the discovery of what the researchers identified as a "perceptual channel across kilometer distances."[12] In other words, the early experiments confirmed that we humans have the capacity to sense and perceive—to remote view—on a nonphysical level that spans across vast distances using an altered state of awareness.

Declassified documents reveal that the early investigations confirmed (1) "that it is possible to obtain significant amounts of descriptive information about remote locations," and (2) "the physical distance separating the subject from the scene to be perceived does not greatly affect the accuracy of perception."[13] During the experiments investigators worked with psychically sensitive individuals to hone and perfect the protocols that produced the best remote viewing outcomes. The results were stunning.

For example, one of the declassified reports describes early investigations to determine the validity of the protocols the teams were developing. The stated objective in one of the projects was to view equipment at the nearby Lawrence Livermore National Laboratory located approximately 50 miles from the viewer at Stanford University. The report revealed that "the overall accuracy of the remote viewing was 77 percent and overall reliability 78 percent."[14] Interestingly, the accuracy of the description of a windmill farm, located just outside of the labs, was 100 percent.[15]

## THE RINGS OF JUPITER

One of the most often reported of the experiments conducted during this time was performed by the well-known psychic, artist, and writer named Ingo Swann. Before NASA's Pioneer 10 space probe was scheduled to pass through the asteroid belt and explore the planet Jupiter, Swann wanted first to remote view the planet himself. He wanted to compare what he saw intuitively, through remote viewing, with the actual images and readings that the spacecraft would eventually send back to Earth.

Swann did so in 1972, and to the surprise of both himself and the investigators, his initial remote viewing detected a ring around what was believed at the time to be a ringless Jupiter. The suspicion of the experts was that he had somehow misdirected his viewing. They assumed he had overshot the target of Jupiter and was actually viewing the known rings on the next farther planet in our solar system, Saturn.

However, the precision and accuracy of Swann's remote viewing skills were confirmed on December 4, 1973. It was then that NASA received the images sent from the Pioneer 10 spacecraft during its Jupiter flyby. From astronomers and geologists to cosmologists and engineers, all were equally surprised. The images clearly showed that Jupiter does, in fact, have a series of rings, echoing what Swann had described in his viewing session.[16] Later studies revealed that Jupiter's rings are made primarily of dust, which is harder to detect from Earth than the ice-based rings of Saturn, explaining why the rings had not been seen by astronomers or spacecraft previously.

## ONCE WE SEE, WE CANNOT UNSEE

The point of sharing the stories in this chapter is twofold. First, I want to demonstrate just how deep our perceptual relationship really is to the field that underlies our reality. Second, I want to show that we appear to be biologically wired to interact with this field through our awareness and intuition in extraordinary, meaningful, and useful ways. The fact that the U.S. military has invested its time, energy, and resources into validating remote viewing, and then developed protocols that could be learned and skills that could be taught to other people, illustrates these points.

Events such as the ones described by the stories in this chapter, and others, give us an indication of the potential that awaits us as we embrace our human potential. They show us that if one person can learn and develop the skills to

- accurately tune in to the well-being of a loved one half a world away;
- elevate their state of conscious to the level that they are no longer confined by physical matter;
- heal life-threatening conditions that conventional medicine is reluctant to address; and
- view distant locations, objects, and people in the past, present, and future and on other worlds with a high degree of accuracy, and do so on demand, then . . .

. . . the evidence suggests that we can all learn to do these things as well. This sampling of awe-inspiring examples reveals capacities and skills that are made possible by our humanness. They're available to each of us. In fact, through some of our most ancient and cherished spiritual traditions, we're told that it is our destiny to embrace these extraordinary abilities, and more.

The method of using one person's extraordinary potential to inspire others to see themselves in a new light was well-known and embraced in the traditions of ancient masters teaching their students. The four gospels of the New Testament, for example, describe the learned master Jesus of Nazareth performing miraculous feats such as the multiplication of limited amounts of food to feed the many, healings of the wounded and sick, and even bringing back to life people who had died.

When asked how he accomplished the healings, Jesus first attributed his seemingly miraculous abilities to his relationship with God, and then he told his disciples that they could do the same as he had done, and even greater things in their lifetimes, if they would allow themselves to believe in what he had shown them and follow the path he'd offered.

For example, in the Book of John (14:12), Jesus says: "He that believeth in me, the works that I do shall he do also; *and greater works than these shall he do*; because I go unto my Father."[17] (author's italics)

These words, although ancient, are not obsolete. When we personally witness the extraordinary feats of others, science has revealed that our bodies—our neural networks, mirror neurons, hormonal secretions, immune responses, and overall physiology—are positively impacted by this experience. The feeling of the impact is our body telling us that what we've witnessed for another person is possible for us too.

## WE NEED REASONS TO BELIEVE IN OURSELVES

The literature of ancient spiritual traditions around the world, such as various Hindu scriptures and biblical texts, tells us that we live in a world of illusion and in an age riddled with deception. The discoveries that I've shared in this book confirm the themes of this literature.

- The *illusion* is that we are flawed, frail, and powerless beings, susceptible to the failure modes of sickness and disease that are part of a world which is separate from us.
- The *deception* is that, in our seemingly vulnerable state of weakness, we need something beyond ourselves to be healthy in our lives and successful in the world.

That savior, we're currently being told, is the replacement of our natural bodies with artificial approximations of our natural biology. The insidious part of the deception is that the solutions that we're being encouraged to accept, in reality, will actually veil the power of our divinity and leave us even more vulnerable to the fear and control that we say we want to heal and transcend.

We owe it to ourselves to preserve and protect the gift of our bodies and the divinity that we were given so very long ago. When we live our divinity in our daily lives, we are reminded that we are a good species by the nature of our humanness.

**PURE HUMAN TRUTH 47:** To live our divinity is to heal the deceptive belief that we are flawed, frail, and powerless beings who need something outside of ourselves to succeed in the world and thrive in life.

When we witness the divinity of others, we awaken to the realization that we are God/eternal within the body. It is this realization that must become the guiding principle of the new world that is emerging and the foundation of the new human story.

CHAPTER SIX

# Human or Hybrid?

## Whose Idea of Progress Are We Following?

We are becoming a hybrid species—
a fusion of biology and technology.

— Dan Brown (1964– ), American novelist

During the 2008 financial crisis, the chief of staff to President Barack Obama, Rahm Emanuel, famously said: "You never let a serious crisis go to waste. And what I mean by that is it's an opportunity to do things you think you could not do before."[1] The statement, and the thinking underlying the statement, became known as Rahm's Rule, and it was used to justify unprecedented actions when it came to financial policies such as cutting interest rates to zero percent and Congress passing the $700-billion Emergency Economic Stabilization Act of 2008, aka the "bank bailout" or "Wall Street bailout," which kept large banks and financial institutions from being liquidated due to the crisis. Not surprisingly, 2008 was not the last time that Rahm's Rule was used during a time of crisis.

In January 2020, World Economic Forum (WEF) founder Klaus Schwab opened the annual meeting of the organization in Davos, Switzerland, by stating that the Covid-19 pandemic presented members with a rare opportunity like no other in history. He argued that the debilitating crisis, including the loss of jobs and businesses, the breakdown of worldwide supply chains, and the collapse of entire economies presented governments and businesses around the world a reason to implement policies and reforms that went beyond simply fixing broken and inefficient systems.

He viewed the pandemic as an opportunity to apply Rahm's Rule on a massive scale—as a vehicle to implement radical changes in global society itself, and ultimately, in all our lives. He named the opportunity the *Great Reset,* a term that is now commonly used to justify the immense social, economic, financial, and human biological changes that are now proposed, and in some cases, being implemented through the United Nations program known as the 17 Sustainable Development Goals. The date by which these goals are to be implemented has been set as year 2030. The movement to implement so much radical change, in such a brief period of time, has triggered much of the economic and social turmoil we see in the world today.

**PURE HUMAN TRUTH 48:** In 2020, the World Economic Forum announced plans for the Great Reset, an unprecedented attempt to build a new digital society in the wake of the global pandemic that was still underway then and the chaos of the lockdowns occurring in many countries that year.

Rather than helping people, businesses, and entire national economies to recover from the debilitating impact of the 2020 pandemic, Schwab's vision was, and continues to be, the remaking of the world around us, and the world within us, in ways that sound like the plot of a dystopian science fiction movie.

During an interview with Schwab conducted at Singapore's Lee Kuan Yew School of Public Policy, he said the world is now experiencing the Fourth Industrial Revolution, or 4IR, for short. For context, the First Industrial Revolution is commonly recognized as the widespread shift, starting around 1760, away from agricultural and toward the industry, manufacturing, and automation that made people's lives much easier.

The Second Industrial Revolution began in 1879 with the invention of the light bulb and a myriad of devices powered by electricity, ranging from assembly lines to build machines and weave textiles, to the invention of plug-in office and home appliances.

The digital revolution that began in 1980 triggered the Third Industrial Revolution. This is marked by the emergence of personal computers and the ownership of data becoming the basis for markets, economies, and manufacturing.

According to Schwab, the Fourth Industrial Revolution is occurring now. It consists of merging all of the technologies from the previous industrial revolutions into a single, highly connected, digitally networked transhumanist reality.

The public proceedings from the WEF presentations reveal that the vision for this vast data sphere is for the entire system to be governed automatically by sophisticated algorithms, deep surveillance, and an advanced system of AI. Schwab and other members of the WEF view the convergence of our physical, digital, and biological lives as the inevitable next stage of progress in the evolution of civilization and humankind.

**PURE HUMAN TRUTH 49:** The vision for the Great Reset is to merge us and all the present-day systems of finance, business, manufacturing, transportation, and food production that we use, as well as our consumption, travel, lifestyle choices, and spending habits, into one vast network that is managed and regulated through the oversight of an advanced AI.

It is my view that the proposed integration of the human body and consciousness into the digital landscape envisioned by the WEF is irresponsible and dangerous, and it poses an irreversible threat to our lives and our humanness.

In June 2019, the WEF and the United Nations signed the Strategic Partnership Framework, a document that outlines a comprehensive set of goals they mutually hope to achieve by year 2030. It's because of the influence wielded by the governments, corporations, and financial institutions that make up the WEF, and this recent collaboration with the United Nations and world governments, that I will use the UN's public statements and declared goals as specific examples of the broader attempt to introduce transhumanism into our lives.

## HACKING US INTO THE MATRIX

Schwab clarified precisely what aspects of our lives will be impacted in the Great Reset. "It's not just a digital revolution. It's digital and, of course, physical with nanotechnology, *but it's also biological.*"[2] (author's italics) It's the biological component Schwab identifies that is most concerning with regard to our humanness. When he says that the digital revolution is "biological," he's saying that part of the revolution includes the merging, or the replacement, of our natural bodies with digital systems for various reasons.

The WEF belief that our evolution as a species is somehow tied to the "progress" of merging nanotechnology and AI with the human body paves the way to a number of unsettling consequences and unanswered questions. In the sections that follow, we'll explore how transhumanists plan to merge technology with the human body and what it means to our everyday lives.

**PURE HUMAN TRUTH 50:** The key to the success of the Great Reset is the digital merging of humans, including our intimate biometrics and vital signs, as well as signs of anger, fear, and joy, into the global matrix to be interpreted by automated systems.

Some elements of the Great Reset are already happening. They're already changing the world around us. These include policies that are replacing small and locally owned organic agriculture with large corporate farms that require pesticides and GMO technologies to sustain themselves, and climate goals that, if they were to be met, would actually create a global environment that we have not seen since the Pleistocene era of geologic time—an era that was not especially good for any life on Earth, including human life. While these elements of the Great Reset are changing the world around us, as we'll see later in this chapter, the biological proposals that Schwab and others have identified are designed to change the world within us.

The most concerning part of the Great Reset is the fact that some aspects of the agenda are already being implemented into our lives and into our bodies without the most foundational questions regarding our relationship to the goals being answered. Or if they are being answered, the answers are not shared openly.

These fundamental questions include: Who decides what technologies are actually good for us? What criteria are used to ensure that they are actually safe? And who is it that determines which populations and groups are be required to adopt specific types of technology? To what degree will we be mandated—and even forced—to comply with those determinations?

It's important to acknowledge that, at the present time, many biology-altering decisions are being made by people and organizations that we, as citizens of our respective countries, did not elect to represent us or our interests. Among those organizations are corporations whose profit-driven missions represent a clear conflict of interest when it comes to making decisions about social policy, medical care, and our lifestyles.

How much sovereignty do we, as citizens, really have when it comes to our choice of accepting or rejecting technology, such as RFID chips (now available as an option for service members in the U.S. military and used commonly in people for business and financial applications in some European countries such as Sweden, and with growing acceptance in France, the UK, and Mexico) or innovative gene therapies and medical technologies, from being placed into our bodies?

If the propositions of the Great Reset simply reflected the casual musings of a few benign, academic intellectuals meeting in a dark and musty old library once a year to discuss world events, I would not be writing about them here. But in fact, these ideas are representative of something much bigger, better organized, and more intentional than that. They directly reflect the intentional actions and declared goals of the WEF, goals that have been honed and refined since Schwab convened its first meeting on January 24, 1971.

## A BRIEF LOOK AT A CONCERNING PARTNERSHIP

In 2019, the United Nations program to modify global society, known as the United Nations Sustainable Development Goals for 2030 (UNSDG-2030), became a legal and political vehicle that could allow for implementation of the WEF's Great Reset. While the 17 social engineering goals of UNSDG-2030 are not themselves the focus of this discussion, they play a role in the threat that transhumanism poses to our humanness and, ultimately, to our innate divinity. For this reason, I'll briefly summarize my concerns using a representative example of the UN Goals.

In 2015, the United Nations proposed this series of 17 goals with the stated purpose being "to transform our world."[3] Further clarifying the intention of the goals, the UN sees them as "a call to action to end poverty and inequality, protect the planet, and ensure that all people enjoy health, justice, and prosperity."[4] At first glance, these goals express a beautiful vision of what could be possible if nations cooperated to make our world a better place. Each one has a title that defines its purpose, such as "good health and well-being," "Zero hunger," and "Sustainable cities and communities." And without a doubt, each represents a heartfelt and seemingly altruistic aspiration. Who wouldn't want good health in a world ravaged by disease? Who wouldn't want to see zero hunger when so much of our global family is experiencing famine?

To be clear, the problem is not in the goals themselves. The problem is in the way that the UN has agreed to achieve them. The 2019 partnership agreement between the WEF and the UN establishes their intentions to leverage the financial and political power of WEF member corporations as the foundation for the Great Reset.

**PURE HUMAN TRUTH 51:** In 2019, the United Nations formally entered a partnership with the World Economic Forum to expedite the implementation of the Great Reset using the 2030 Sustainable Development Goals as the vehicle to do so.

For example, Sustainable Development Goal 2 (SDG 2), the goal of "Zero hunger," is one I'm particularly passionate about. Global statistics show that today, we currently have more than enough food to feed every child, woman, and man on the face of the Earth. There is no supply shortage. It's not because of a lack of food that members of our global family are hungry and starving. The problem is a lack of leadership that would make it a priority for the food we already have to get to the places and people that need it most.

When I mention leadership, I'm not necessarily talking about the political leadership of a nation like the United States, the United Kingdom, Russia, or China. I'm talking about the thinking of all leaders, including you and me, and the way we lead in our communities with our policies and politics. Few leaders have yet to make universal human well-being a vital priority. With these factors in mind, I would think that the delivery of existing food supplies to places of the greatest need would be a stated goal and priority in the goal of "Zero hunger." Yet the details of the plan for SDG 2 are focused upon the technology and business of agriculture, rather than this vital shift in thinking and human priorities.

SDG 2 proposes to "increase investment, including through enhanced international cooperation, in rural infrastructure, agricultural research and extension services, *technology development and plant and livestock gene banks in order to enhance agricultural productive capacity* in developing countries, in particular in the least developed countries." (author's italics)[5]

With the vision of both the WEF and the UN clearly stated, the direction for this goal is telling us that the absorption of small, private family farms into large corporate conglomerates, utilizing genetic engineering for the crops themselves, as well as for modifying the biology of insects that are drawn to the crops, along with application of specialized chemical additives and pesticides, are part of the vision to achieve the goal of eliminating hunger worldwide.

This theme of replacing local methods of farming that families and communities have traditionally relied upon with centralized programs run by big business and high-tech firms is not isolated to only one of the goals. For example, the way to achieve Sustainable Development Goal 3, which aims "to ensure healthy lives and promote well-being for all at all ages," includes policies to "support the research and development of vaccines and medicines for the communicable and noncommunicable diseases that primarily affect developing countries, [and] provide access to affordable essential medicines and vaccines . . ."[6]

While there is a time and a place for some of the traditional vaccines that we've used successfully for over a century for vulnerable populations, the generalized use of new gene therapies proposed to meet the goals for these programs—therapies that are so new they lack the knowledge of possible long-term and lasting effects on children as well as adult populations—may be more problematic than the diseases themselves.

I'm sharing these examples to provide an idea of how what used to be the isolated and utopian vision of a few individuals meeting each year in Davos, Switzerland, has now merged with the efforts of the globally supported and funded staff of the United Nations, and how their combined focus has the potential to play a leading and concerning role in our lives.

Together, their combined vision won the approval of over half of the UN member nations when it came to an up-down vote on whether to implement SDG-2030. This internationally ratified document now allows for a digitized, surveilled, and controlled system of transhumanism and governance to become a reality in our lifetimes.

In the following sections of this chapter, we'll take a deeper look into the proposed policies, the technology that will make them possible, and what they could mean to us, our evolution as a species, and our innate divinity.

## SETTING THE STAGE FOR THE GREAT RESET

In 2018, a full two years before the 2020 pandemic that triggered the Great Reset, Yuval Noah Harari, an influential voice in WEF policymaking, presented at Davos a dystopian vision for the world and our place in it that he believes awaits us only a few short years from now. Harari is a prolific author, a philosopher, and a professor of history. A 2022 article published in *Current Affairs* singled him out as the advisor and consultant of choice for "some of the most powerful people in the world."[7] Those he has advised include members of the United Nations Educational Scientific and Cultural Organization (UNESCO), executives at the International Monetary Fund (IMF), and media influencers such as Meta founder Mark Zuckerberg.

Harari's visions of our future were first featured at the annual WEF meeting in 2018 and then again during the height of the pandemic in 2020. It's because of his influential voice and the magnitude of the changes that the WEF and UN are attempting to impose upon global society that his vision is particularly troubling.

Harari began his 2020 presentation by authoritatively and matter-of-factly stating that ours is "probably one of the last generations of *Homo sapiens*."[8] He then said that within a few short years, the rapid advances in technology will make it possible for us to engineer "bodies, brains, and minds" to the degree that we will become a species more different from ourselves at that future time than we are today from the Neanderthals that lived 40,000 years before us.

The one factor that will allow for such a drastic shift to occur so quickly, Harari says, is that human biology is now recognized as a hackable "technology." He believes that the complexity of our bodies can be reduced to a system of algorithms that can be known, revised, and ultimately rewritten.[9] But just as we discovered in the example of the gene-edited twin girls in Chapter Three, it has yet to be determined whose idea of "better" or what kind of perfection such an edited human will conform to.

As the familiar adage that *beauty is in the eye of the beholder* reminds us, the arbitrary idea of what makes a better or more perfect human appears to be in the eyes of those doing the gene editing. It's because of this godlike ability to rewrite the genetic code of life itself that Harari believes humans will be modified to interface with the digital world in ways that seemed unthinkable only a few short years ago.

In his Davos presentation, Harari elaborated on his perspective of our future, acknowledging that for the four billion years that have preceded us, all life has been governed by two factors:

- The laws of natural selection
- The principles of organic biochemistry

The technological advances available to us today now make it possible to short-circuit both of these factors. Or so the proponents of the Great Reset believe. They envision the forms of life in the future as being defined not by the natural evolution of biological life, but as products of our intelligent design. "Not intelligent design of some god above the clouds," Harari clarifies, "but our intelligent design and the intelligent design of the clouds: the IBM cloud and the Microsoft cloud."[10]

Harari summarizes this portion of his 2020 presentation by stating that we are now breaking out of the limits that have confined organisms to natural organic life and entering an era of life that includes inorganics. In other words, he believes that we're at the precipice of developing forms of life that are machine-based rather than carbon-based as we are. The problem with this line of thinking when it comes to humans stems from the thinking described earlier in this book. The evidence described in Chapter Three suggests that the consciousness that animates a human body, and is the source of our empathy, conscience, and intellect, is not found within the body itself. Rather, the cells and genetic material that make up the physical body are the conduit to a part of us that is not contained in the physical body.

**PURE HUMAN TRUTH 52:** Speaking at Davos to the organizations, corporations, and financial institutions whose operations and purposes impact our daily lives, Yuval Noah Harari declared that human biology is now a hackable "technology."

When I hear the vision that Harari shares, my reaction is immediate and definitive. Immediately my gut says, "No!" No to the trajectory that we have placed ourselves upon. No to normalizing the acceptance of technological gadgets into our bodies on a wide scale. And no to the ease with which some futurists accept this technological direction as progress and an inevitable next step in human evolution. If Harari was the only person identifying this possible future, I would probably not even mention his vision here. But Harari is not alone.

Echoing Harari's views, Amy Webb, a professor of business at NYU Stern School of Business, elaborated upon what the rewriting of various species' genetic codes means. During a special WEF panel discussion, she said, "We're talking about improving biology and redefining organisms for beneficial purposes. It's going to allow us to not just edit genomes, but also and importantly, [to] write a new code for life. We'll have write-level permission."[11]

The *write-level permission* that Webb mentions is a reference to computer programming lingo that means the programmer has the ability (the permission) to actually access building blocks of a code at the most basic level. It's this access that makes it possible to shift the blocks, rewrite the code, and produce new outcomes.

The same computer programming principle applies to us when it comes to permissions to intervene in the fundamental DNA code of human life.

Having reached the moment in history when we have the ability to delete, cut, paste, splice, and silence (inactivate) the genes that make us who and what we are means that just as it is with access to the computer's code, we can be altered, modified, changed, and remade in ways that could make us become something very different from what we've been in the past. It also means that an entirely new form of human life can be created with characteristics and capabilities that separate it and make it very different from other members of our kind.

The applications of such new life-forms could range from human-like beings resistant to toxins and radiation performing tasks that would harm pure humans who are not enhanced, to the development of super soldiers with "biologically enhanced capabilities" that the Chinese military is reportedly undertaking.[12]

The fact that this kind of human modification is even being considered reflects a philosophy that is flawed to begin with. That thinking is the premise that something within us is broken and needs "fixing."

## FIXING HUMAN "FLAWS"

We live at a time in history when the exceptional nature of our humanness is being questioned, denied, and reduced to a sequence of variables in the equation of life. As variables, these qualities can be manipulated, revised, and even replaced altogether. In a paper titled "Transhumanism and the Death of Human Exceptionalism," writer Peter Clarke encapsulates this thinking, stating, "As a philosophical movement, transhumanism advocates for improving humanity through genetic modifications and technological augmentations, based upon the position that there is nothing particularly sacred about the human condition. It acknowledges up front that our bodies and minds are riddled with flaws that not only can but *should* be fixed."[13]

British philosopher and transhumanist David Pearce has further emphasized the so-called flaws of our humanness. During an interview that was conducted with him and a fellow co-founder of Human+ (formerly known as the World Transhumanist Association),

Pearce asserted that the use of technology is the only option we have to achieve the world of peace and well-being that we long for, saying, "If we want to live in paradise, we will have to engineer it ourselves. If we want eternal life, then we'll need to rewrite our bug-ridden genetic code and become god-like."[14]

So, what are the "bugs" in our humanness that need to be fixed? It may come as no surprise at this point that some of our most cherished qualities that we value and celebrate are among the perceived flaws to which Pearce was referring.

In the following sections, I'll identify some of the key characteristics that transhumanists view as human flaws.

## Flaw 1. The "Flaw" of Human Emotion

One aspect of our humanness that is most often referred to as a flaw is our capacity for deep, and sometimes seemingly illogical, emotion. From the violent power to hurt and destroy one another to the miraculous power of healing that stems from our joy and our love, human emotion remains as an uncertain and mysterious force. To date, attempts to engineer and replicate human emotion using software have been met with frustration and disappointment.

Entire careers and lifetimes have been dedicated to cultivating and developing the expressions of our emotions that we celebrate as music, poetry, sculpture, painting, and dance. Creators do so in the hope that their shared experiences will somehow quench the thirst for the connection that we all long for. It's our individual emotions that form the collective bonds that unite us as communities and societies.

Writer Jonathan Cook describes the thinking that proposes to dismantle human emotion. "While some versions of transhumanism envision an enhanced emotional life enabled by integration with digital technology, many of the foot soldiers currently seeking to enhance human life with emotional AI are moving in the other direction, *using sentiment analysis tools to teach human beings how to repress their emotions*."[15] (authors italics)

From my perspective, an AI that can somehow moderate and ultimately repress human emotion is the last thing we need. Developing it is possibly one of the most dangerous paths we could ever choose to follow as a species. In our era of extremes and with conflict running rampant throughout the world, emotionless beings would be powerless to empathize with those suffering the most.

Our ability to feel and emote what we feel is perhaps one of the most powerful, as well as one of the most hopeful and promising forces in our lives. It's the power of our emotions that enables us to appreciate the beauty of an evening sunset on a beach, a morning walk in a growing forest, and the sexual chemistry that magnetically attracts us to others. Our emotions also have a direct relationship to the empathy that we may have for others, including our enemies. It's our empathy for the lives, family, and suffering of those we find ourselves in conflict with that often leads to the negotiations that make peace possible.

Recent discoveries in modern science now add to a growing body of evidence suggesting that our ability to experience emotion in the presence of beauty is more than simply a pleasurable experience; it's a very real and transformative power. Our experience of beauty, for example, and the emotion that beauty elicits within us, is a direct, sensual, and life-altering experience.

We humans are believed to be the only species of life with the capacity to perceive beauty in the world around us, as well as to seek it out in our everyday lives. It's through our experience of beauty that we're given the power to change the sensations we feel in our bodies. Our feelings, in their turn, are directly linked to the way neurons throughout our body "wire" (connect) and "fire" (activate), thus altering the chemistry of our cells and organs. Beauty has the power to change our individual lives, and it's no exaggeration to say that the same beauty that changes our lives has the power to change our world!

The key to allowing emotions to alter our perceptions and behavior is that we must choose to look beyond the hurt, pain, and suffering that we're presented with in the moment to recognize the beauty that already exists in all things. Only then will we have unleashed the power that the choice of beauty holds for our lives. And to do so requires the evolution and honing of the power of emotion. Cook sums this up beautifully. "In order to be worth living in, the future needs to be more, rather than less, emotional."[16]

## Flaw 2. The "Flaw" of Birth from a Womb

Since the time of our origin, women have grown our babies inside of their wombs and risked the potential consequences and pain of childbirth in exchange for the benefits of bringing new life into the world. Natural childbirth can be beautiful and joyous. Some women orgasm as it is happening. It can also be bloody, painful, physically injurious, and full of uncertainties and complications, including in some cases being fatal to mothers and babies.

According to Austrian psychoanalyst Otto Rank, the baby's experience of birth and the separation that follows live birth, including being suddenly thrust from the security of a warm, liquid womb into a cooler, drier environment and the physical shock of severing the umbilical cord, is the first, and perhaps the deepest, trauma of our lifetimes. He believes this experience sets the course of our emotional well-being for the rest of our lives.

Transhumanists now see the trauma, suffering, and associated dangers of live birth as obsolete and optional. They have their eyes set on "fixing" the flaw of women gestating babies inside their bodies through the use of advanced gene-editing technology and artificial wombs, and doing so by the year 2050.

We're actually closer to the reality of nonuterine births than many people realize. In a study published in the peer-reviewed journal *Nature Communications*, Emily Partridge of Children's Hospital of Philadelphia led a team that successfully created life-giving, liquid-filled artificial wombs for eight lamb fetuses. Sheep are often used as substitutes for humans in research due to similarity in veins, arteries, and brain size.[17]

The lambs in this study were between 105 and 115 days old, the equivalent of a 23-week-old human fetus, which is a key age for developments such as REM sleep and responses to the environment. The artificial wombs allowed the lamb's brains, organs, and even their fur to develop normally during their time spent inside the devices.

If the experiments continue to be successful, and the advanced technology is approved by the FDA, as is expected in the near future, the researchers believe that the artificial wombs will be used on human babies under special circumstances within three to five years.

While this is welcome news for doctors and mothers facing medical extremes that might lead to dangerous, and sometimes fatal, premature births, once the artificial womb option is in place, transhumanists believe the new devices and the gene editing of embryos will soon become commonplace. These technologies, they believe, will replace the uncertainties of natural conception and the possible complications of natural childbirth.

A leading voice in the transhumanist movement, Zoltan Istvan, speculates on the timeline for achieving human birth without a uterus: "The current goal of the growing field of ectogenesis birth [author's note: birth without a natural womb] is to use it to keep human premature babies born around 23 or 24 weeks alive—and still growing. Eventually, these advances will likely lead to an era of children born without the need of women's uteri at all."[18]

Historically, the act of human conception—sex—and the decision to bring a baby into the world has been intimately linked with bonding between partners and the structure of a family. This bonding includes the implied commitment that the resulting baby will be nurtured and provided with the skills needed to thrive in its immediate environment. One of the possible consequences of transhumanist births using stand-alone artificial wombs, says Istavan, is with regard to these factors. The ability to conceive and give birth without a woman's womb "could eventually lead to a society where relationships, sexual or otherwise, are not functionally necessary to continue the human species."[19]

In other words, in light of these technologies, it's not so far-fetched to envision a world of baby-producing factories consisting of large warehouse-like structures where rows of artificial wombs contain living human fetuses. During their development the fetuses might be "optimized" for specific functions such as athletic performance or increased IQ—or to become super soldiers—with gene-editing technology, and monitored by computers and AI until the moment of their . . . *birth*? We don't, as yet, even have the vocabulary in place to name the outcome of such a drastic shift in human life. It wouldn't be accurate to say that these babies are "born," as there would be no actual delivery as typically occurs with the emergence through the birth canal.

At the end of their development in the artificial womb, the babies would literally be *transferred* from the liquid environment of the technology into the dry air and waiting hands of a technician, or into the arms of a mechanized receiver—a robot. If this sounds like something from the 1999 sci-fi movie *The Matrix*, we have to ask ourselves: Where do the ideas for such futuristic visions come from?

The current thinking of some researchers and futurists is that the movies we create today are the reflections of mass consciousness informing us of what we may expect to experience in our lives based upon the current trajectory of policies, thinking, and technology. In today's world, we're seeing the precursors of precisely the outcome that I've just described, beginning with the rapid and unprecedented decline in global birthrates.

According to United Nations data, the world reached a peak level for women's fertility in 1963, with an average of 5.3 children being birthed that year per woman.[20] Since that time, this rate has declined over 50 percent to an average of 2.32 births per woman.[21] As of 2021, over half of the countries of the world are now at a birthrate that is actually below the replacement rate of 2.1 children per adult woman needed just to maintain the current global population.[22]

Human fertility is complex, and there are a number of factors contributing to the lower birthrates. These include toxins in the environment and hormonal disruptors that affect the quality of men's sperm as well as the viability of women's eggs. Additionally, social and economic factors in wealthier countries have now shifted the focus and the meaning of growing a family to the point that it is viewed as optional rather than necessary.

The point I'm making here is that the very real and rapid decline in fertility, coupled with the option to bring a baby into the world without the physical pain and suffering of doing so, has opened the door to wombless births producing "designer gene" babies—the transhumanists' vision of engineering away human flaws to create the ideal society.

Because transhumanists view some of our most valued characteristics as flaws, the ability to "edit" away our cherished human capabilities is particularly disturbing. This is precisely what Yuval Noah Harari was referring to when he stated that we are among the last generations of *Homo sapiens*. He was telling us that the technological advances that we have now, the thinking underlying the direction we're headed, and the outcomes that are already emerging are leading us to becoming hybrid forms of life that are no longer purely human.

**PURE HUMAN TRUTH 53:** Transhumanists view some of our most cherished characteristics and abilities as flaws that can be remedied and "fixed" using the advanced technology of gene editing and nanorobotics available today.

It's this new species of techno-human hybrid that is necessary in order to implement the principles of the Great Reset. Stated another way, in the absence of the techno-human hybrids that are proposed, the Great Reset cannot be implemented. Once the flaws of emotion and imprecise outcomes from random conception have been "fixed," the transhumanists believe that the next step is to fix the "flaws" in society.

## WHAT WOULD A TRANSHUMANIST SOCIETY LOOK LIKE?

The proposed Great Reset world that would "fix" our postpandemic society is a world of replacement. It's a world that replaces our ideas and values, such as individuality, uniqueness, innovation, and self-reliance, with a muted and homogenous society that is viewed as fair and equal for all. The problem is that what's thought of as fair and equal by some, particularly those in power, is a world in which others of us have lost many of our hard-won freedoms.

It's a world where speech, ideas, and information are controlled. They have to be so that new innovations and options that threaten the status quo are not known or made available. Rather than the traditional and time-honored world of localized living that has proven to be healthy, sustainable, and viable for millennia, the Great Reset world is a world of centralized power, centralized control, centralized resources, and sameness, all regulated using a new wave of surveillance technology and "smart" systems.

In his 2020 talk to attendees of Davos, Harari described how the AI algorithms of the Great Reset will know us better than we know ourselves. They will predict everything from our future health conditions and lifespans to knowing our sexual preferences even before we identify them for ourselves as children. Harari shares from his own personal experience how different (and presumably better) his life would have been if he had known early on what his sexual feelings meant, rather than having to go through the struggle he did to arrive at the conclusion of same-sex preference that an AI could have offered him years earlier.

The algorithms, he argues, will know these things about us because of physiologically embedded, digital human-machine interfaces that will allow them access to the most intimate details of our lives. This will include access to our innermost thoughts and feelings about our personal life experiences.

Such transparency may enable AI to help us and make aspects of our lives more convenient, but it can also present the proverbial double-edged sword. It could be possible, for example, for the truth revealed by our biometric data, such as our heart rate, our galvanic skin response, and our pupil dilation, to reveal our authentic feelings about a new government policy or a healthcare mandate with which we don't agree—a truth that could potentially be used against us. The data would correctly show that we emotionally rejected the new policy, even if outwardly we nodded our heads, clapped our hands, and did things to support it.

Hopefully, this level of data abuse never happens. The point here is that the technology is already available to support such a scenario if the choice is made to do so.

This is an example of why the rapid onset of merging technology with our bodies and the attempt to normalize the use of gadgets inside our bodies is so concerning. While today's nanodevices are certainly the most advanced we've seen in human history, and can certainly be applied for the greatest benefits of health and healing, they're being developed in an environment where the powers that be are jockeying for leverage and control of them in an attempt to "perfect" our species.

We've seen this before. What we're witnessing today is just the most recent iteration of an old idea that first gained momentum early in the 20th century. At that time, it was called *eugenics*.

## A NEW NAME FOR AN OLD IDEA

A dictionary definition of *eugenics* is "the practice or advocacy of controlled selective breeding of human populations (as by sterilization) to improve the populations' genetic composition."[23]

The CRISPR gene-editing technology, described previously, now makes it possible to choose the characteristics of a human baby before it's born. The successful births of the gene-edited twin girls in China demonstrated that this technology is already beyond the level of theory. The moral and social implications of gene editing humans have placed us in uncharted territory and reignited a conversation regarding this sort of technology and what it means in our lives. Although the resurgence of the conversation may seem new to members of younger generations, the idea of stacking the deck of natural selection in favor of certain characteristics actually began long ago.

In 19th-century Europe, for example, it was common to encourage families that had "desirable" traits, such as beauty, high intelligence, and good health, to have as many children as possible in an effort to infuse the local population with those good traits. The dark side of this thinking is that individuals who had traits deemed undesirable, such as a criminal history, mental illness, physical disabilities, or low intelligence, for example, were discouraged, and often prevented, from giving birth. The practices used at the time to prevent those births included forced sterilization, primitive forms of abortion, and the legal ability to prevent certain marriages where the outcome of a pregnancy would be suspect. These practices became formalized and institutionalized early in the 20th century under the umbrella term of *eugenics*.

The obvious problem underlying the philosophy of eugenics is with regard to who is setting the criteria and making decisions that determine which human qualities are "good" and which are not. In times past, these decisions, the racism, and the horrible consequences of such decisions were often based on pseudoscience and public opinion.

**PURE HUMAN TRUTH 54:** Using gene editing and chemical therapies to manage human conception and births to ensure a predetermined outcome of "desirable" traits is the high-tech version of the old philosophy of eugenics.

A 1999 paper published by *BMJ* presents three concerns that arise from eugenics thinking, and does so from a scientific rather than a purely emotional perspective. The paper states: "The most common arguments against any attempt to either avoid a trait through germline genetic engineering or to create more children with desired traits fall into three categories: worries about the presence of force or *compulsion*, the imposition of arbitrary standards of *perfection*, or *inequities* that might arise from allowing the practice of eugenic choice."[24] Following is a brief discussion of each of these concerns.

## Concern 1. Compulsion

We need look no further than the eugenics movement of the early 20th century for examples of how a society can impose—compel—restrictions of reproductive choices in an attempt to limit, or even eliminate, entire portions of a society. Between approximately 1907 and the early 1970s, over 60,000 people in the United States were legally sterilized against their will, and it was done under the auspices of the eugenics legislation that was accepted at the time.[25] The *BMJ* article, cited in the previous section, states that compulsory reproductive policies of any kind, including forced sterilization, reproductive habits, and mandated partners, is "morally objectionable" from any form of authority, including governments and institutions.

The right to be free of such limits is so broadly acknowledged and accepted in the international community that it is foundational in international laws. The article also acknowledges, however, that it is possible to embrace technology and practices in order to pursue a desired birth outcome of health or genetic advantage with informed consent, by choice, without being compelled to do so.

## Concern 2. Arbitrary Standards of Perfection

The transhumanistic objective of optimizing human qualities is deeply rooted in the idea of achieving "perfection." But the concept of human perfection is subjective at best. Unless the perfection that is being sought is with regard to physical health, healing, and mental and emotional well-being, the criteria for perfectness is typically determined by culture, society, and environment. It can change over time. The criteria is determined by comparing an individual, or an entire class of people, to some kind of arbitrary standard that has been agreed to by those doing the comparison.

The 1999 *BMJ* article makes a distinction between seeking to determine the subjective traits of a child, such as hair color and eye color, which may be a matter of the personal preference for the parents of the child, and population eugenics, where the ideas of an elite few are forced upon an entire population in an attempt to "weed out" characteristics that have been deemed undesirable by them.

The world witnessed an extreme example of this kind of eugenics between 1933 and 1945, when Nazi Germany used the forced sterilization of over 400,000 people, selective breeding, and ultimately, mass euthanasia to cleanse the state of ethnic minorities, LGBTQ-plus populations, and people who were subjectively determined by the authorities to be unworthy of life, like individuals with epilepsy or Down syndrome.[26]

While the hope is that we have learned from the atrocities of the 20th century and recognized just how dangerous eugenics thinking is, the transhumanist movement and the technology that makes the human-digital possible has reignited the conversation. In fulfilling the vision of the Great Reset, whose idea of perfection will be used as the benchmark of perfection? Who gets to decide what qualities are worth preserving when it comes to engineering a child's life and choosing the characteristics that will guide that life in the world?

## Concern 3. Inequity

In addition to the concerns of compulsory reproduction and the subjective nature of using gene editing to produce "perfect" babies defined according to arbitrary standards, there is the very real possibility, perhaps even a probability, of producing a stratified society consisting of those who are optimized and those who are not.

This caste type of society could easily result from the already-present economic factors that separate families in our societies around the world. Those parents who could afford the medical care and had access to the technological advances that allowed them to choose the characteristics of their children would obviously have an advantage over those who could not afford care and did not have access to such options.

We could easily end up with a world of technological haves and have-nots, where the engineered athletic and cognitive abilities and physical characteristics will be used to create a social hierarchy that dwarfs the economic and educational inequality that we have today. The argument can be reasonably made that such a hierarchy already exists in some places today. It's the factors described in the previous sections that stand to deepen the divide by adding to the factors that create social privilege.

• • •

The concerns identified in the 1999 *BMJ* article are real, and they're worrisome. In all probability, the repercussions of all three of the issues associated with eugenics that it mentions—the presence of compulsion, arbitrary standards of perfection, and inequity—are likely to continue for the foreseeable future. They are likely to do so because they reflect a current and growing way of thinking in the world today. Although the concerns are given modern names and accepted as the product of "tradition" or culture, we already have the principle of *compulsion*, for example, as arranged and forced marriages to keep property and bloodlines intact.

We already have the principle of *perfection*, expressed in the subjective ideas of beauty in our society and in an entire industry of cosmetic surgery to accommodate the current thinking about what's desirable when it comes to aging faces and physical body proportions, for example. We already have the reality of *inequity*, expressed in the lives of the economic haves and have-nots who can and cannot benefit from the advances in technology and the selective gene editing of their babies before they're born.

The new concern is that rapid advances in technology are now making it possible to quickly spread this transhumanistic thinking from the few industrialized nations where it is being pioneered, and spread it on a global scale. If this were to happen, the empathy we aspire to feel for one another, and the intimate relationships that we cherish as a species, would quickly become remnants of our past.

**PURE HUMAN TRUTH 55:** The philosophy of eugenics and selective breeding is inherent in the transhumanistic vision of the Great Reset.

Perhaps the saddest part of this transhumanistic scenario is that, after the first generation of modified humans is indoctrinated into this way of thinking and living, and the use of nanotech-monitoring devices has become accepted and normalized in our everyday lives, it's all the world will know. The new generation will know only the experience embraced during their lifetimes. Human sovereignty and the idea of personal privacy on an intimate level will become obsolete remnants of an outdated past.

In many respects, we're already seeing this happen today. And the thinking that makes this kind of indoctrination possible is the consequence of recent generations being given no reason to honor their bodies or respect the gift of their humanness.

## WORSHIPING TECHNOLOGY

We live in an age where advanced technology has developed so quickly that there are entire demographics of young people who cannot imagine a world without it. They cannot imagine a world, for example, where they don't have immediate access to their computers, iPads, smartphones, the Internet, and connection to their friends. They can't imagine it because they've never lived in a world without these things.

For example, people who were born between approximately 1982 and 1994, give or take a year or so, fall into a demographic known as Generation Y, Gen Y, or more commonly, the Millennials. Millennials can still remember the world *before* cell phones burst onto the scene and computers became necessities for modern life. They can remember when computers were still considered optional business machines used primarily at work to summarize statistics on spreadsheets.

It was only with the development of microprocessors that made smaller machines possible in the 1990s that personal computers became commonplace in our homes. Even then they were used more for playing video games and watching movies than for business applications. The generations that have been born since the mid-1990s have lived in a world where advanced technological devices have always existed. They've never known a world where they couldn't hold a smart device in the palm of their hand to find the answer to any question of history and science, or to listen to any music from any artist or composer, from any time in history, or instantly calculate how much of a tip to leave with their waiter for lunch.

This generation has also never lived in a world where the personal lives of friends, family, and celebrities wasn't shared across multiple platforms on social media. It's precisely this generation that has been targeted, and successfully indoctrinated through our public education systems, to view themselves as inferior to the technology that they rely on to do things. And it's for this reason that the generations following Gen Z are now in danger of succumbing to the philosophy of the digital eugenics that now threatens our species.

They are vulnerable because the authority figures they trust today, and that they've trusted in the past, as well as the teachers and professors that we've trusted to educate our children, with rare exceptions, have never given them any reason to think about themselves in good and positive ways. It's not that their unique qualities and human specialness have simply been left out of their indoctrination. Rather, it's that there's been a concerted effort to prevent our young people from recognizing these qualities of their specialness and accepting the power of their divinity.

**PURE HUMAN TRUTH 56:** Younger generations are especially vulnerable to the threats of transhumanism because they've grown up in a technological world where the lure of computers and AI is touted as humanity's saving grace.

Young people today have been conditioned to see themselves as powerless victims of a world over which they have no control. They've been taught to fear and hate carbon, for example—the very stuff that their bodies, all life, and our planet is made of. They've been taught that carbon is bad, that it's a poison to our world, and that anything made of carbon and that anything that uses carbon is flawed by its very nature.

It's this thinking that leads to a victim/savior mentality. And this is precisely why the allure of transhumanism, and the underlying eugenics philosophy, may seem attractive to younger generations.

Who wouldn't want a computer chip in their brain that uses less carbon energy and allows them to play games like *Grand Theft Auto* and *Call of Duty* on their computers without needing a cable to connect to a machine? Who wouldn't want an AI to compose their term papers and science reports when the AI on their laptops is faster and more efficient in doing so than they are?

Why wouldn't they want ChatGPT to compose music to go with the lyrics of the song that the same ChatGPT composed only moments ago? Or to create an exquisite image on a canvas in only minutes that used to take hours for a human to produce with inspiration, skill, messy paint, and the possibility of producing something less than a stellar piece of art?

The answer to each of these questions is the same. Each time we give away our creativity, imagination, and skills of reason to technology, we lose something. We lose a piece of ourselves. We lose a portion of our humanness. And while we may not recognize that this is happening, and the immediate impact may seem small at first, there are deep, cumulative, and lasting effects that we owe it to ourselves to seriously consider.

It's precisely the loss of our humanness and divinity that is given as the reason for some of the most mysterious and unexplained events that have swept the world since the 1960s.

## A WARNING TO REJECT TRANSHUMANISM

In the early 1990s, Harvard Medical School professor John Mack began a project to systematically document the myriad reports of people being temporarily taken against their will by what appear to be otherworldly beings navigating futuristic craft that are not human-made—the phenomenon of *alien abduction*. Mack made headlines with his reporting, as he was a Pulitzer Prize–winning author, physician, and head of the university's department of psychiatry from 1977 until his mysterious death in 2004.

Mack endured criticism from the Harvard administration, public humiliation, and was even shunned by some of his peers for his attempts to formally investigate the alien-abduction phenomenon as a scientist using scientific tools and methodologies. The criticism stated that by conducting a scientific investigation, he was legitimizing a phenomenon that was commonly thought to be the result of emotional issues with the people reporting abductions and something that academia preferred to distance itself from.

The thinking at the time was that the people claiming to have been abducted were the victims of unresolved psychological issues, schizophrenic, or simply "crazy." One of the key takeaways from Mack's work, however, was that through his personal examinations of hundreds of abductees, he demonstrated that they were definitely not crazy. Though many were shaken up by their experiences, as would be expected, and were unable to make sense of what had happened to them, they were demonstrably and clinically sane. The consistency of details in the abductees' different accounts of their experiences, and their ability to recount what happened in a rational and methodical way, meant that there was something else happening. Whatever that "something" was, mainstream academia was overlooking it.

I had the opportunity and good fortune to know John Mack personally and tour with him on the conference and seminar circuit in the mid-1990s. It was through those meetings, our conversations, and reading similar reports from other researchers regarding alien abduction that I became aware of how widespread the phenomenon really is. It was also during this time that Mack was suddenly killed one evening after speaking at a conference in London.

Following an evening dinner with friends, he was walking home alone along a London street. Suddenly, at 11:25 P.M., a car reportedly driven by a drunk driver left the road, drove up over the curb, and struck Mack as he was walking. The driver then returned his vehicle to the street, where he was arrested for the crime. Mack lost consciousness at the scene and succumbed to his injuries shortly after.

The controversy surrounding Mack's work and the circumstance of his untimely passing continues to this day. Fortunately, Mack succeeded in publishing his landmark book detailing his research, titled *Abductions: Human Encounters with Aliens*, before his death.

The publication of Mack's book opened the door for other people who had been reluctant to report similar experiences to come forward and tell their stories. And there were not just a few isolated cases. Suddenly, there was a global outpouring of people saying that they had encountered beings from another world, been taken against their will to places they believed were inside moving aircrafts, been allowed to look through portals outward into space and the stars, been intrusively examined and then returned to their cars, homes, and bedrooms by their abductors.

The common thread to the myriad abductions is what the abductees were told when they asked their captors obvious questions like: Why had they been chosen? What was the purpose behind their abduction? What were the examinations all about?

For some of the abductees, the answers they received to these questions were as unsettling as the abductions themselves. While the exact wording of the answers they received varies from incident to incident, the themes that they describe do not. Based on both telepathic communications and direct verbal communications, the abductees describe their captors as advanced beings that have traveled to Earth from distant worlds. They say these beings are truly alien to us and have come to Earth for one purpose: *to warn us about the technological choices we are making in our lives and in our bodies, and the consequences our choices may lead to for us as a species and as a planet.*

**PURE HUMAN TRUTH 57:** A common thread that is documented among the reports of alien abductions throughout the world is that the abductors are warning us of the consequences of choosing a transhumanistic path of human evolution.

The warnings are typically focused upon the way we're using our advances in technology, and the dark consequences that await us if we continue upon our current path. The warnings include the possibility of global atomic war, an ensuing nuclear fallout, and the destruction of our global climate, diverse ecosystems, civilization, and ultimately, our species, if such an event were to occur.

The warnings are also directed toward us and our impending choice to merge computer chips, sensors, and synthetics with the human body—transhumanism—and the potential loss of the very characteristics that make us such a unique form of life. Ultimately, the warnings reflect the concern that, if left unchecked, we are on a destructive path that will forever degrade, and possibly end the human species as we know it.

## WE HAVE WHAT THEY'VE LOST (AND WANT BACK)

The beings from faraway worlds have an interest in us here on Earth because of where we are in our human evolutionary process. They tell abductees that we are at a critical crossroads in our world today, one that the beings themselves also had to pass through at some point long ago, in their ancient history. They had to make a choice between continuing their species as a natural form of biology or replacing their natural bodies with artificial technology—and this choice would forever define their species and its future evolution.

To be clear, they had to choose between merging their natural bodies with computers and AI and developing the natural abilities associated with their neurons, soft tissue, and capacities for self-regulation. The choice they're referring to is precisely where we stand today that is the subject of this book. We've reached the time in our evolutionary journey when we must make the choice of how much technology we allow to become part of our physical bodies. Another way of saying this is that, as a species, we are collectively determining how much of ourselves we will give away to the technology that we're being encouraged to accept today.

The catch the aliens want us to be aware of is that once we make the decision to become a hybrid species, we cannot go back. Our choice will be permanent. We can never return to the form of life that we once were because of the way our bodies will adapt to and change in the presence of the artificial devices we accept and implant in them. Eventually our cells, including our neurons, and entire systems, such as the immune system, will no longer function as they have in the past due to lack of use. Ultimately, these abilities will atrophy and disappear forever. In this way, we lose the qualities that set us apart from other forms of life.

And this is the key to understanding the alien abductions. The abductors are interested in us because in their past they chose to follow the path of technology. They chose to replace their natural bodies with machines and gadgets. And while their computers and AI have made them fast and efficient, they have lost abilities and characteristics they used to have. Now, they want their prehybrid lives back. Their hope is that the splicing of pure human DNA and human tissues extracted during the examinations of abductees into their own bodies will make it possible for them to restore at least a portion of their original biology.

## NEW MEANING FOR ABDUCTION STORIES

On July 26, 2023, former intelligence official and now whistleblower David Grusch presented testimony, under oath, before the U.S. House Committee on Oversight and Accountability.[27] The focus of his testimony was regarding his knowledge of secret projects that are actively hiding the realities of crashed and captured UFOs. In addition to the recovered technological remains of such craft, Grusch described some of the advanced, reverse-engineered technologies that are now in the possession of the United States military.

Trying to clarify precisely what was recovered with the advanced craft, South Carolina's Representative Nancy Mace asked Grusch if we also had the bodies of the beings that had piloted the craft in our possession. Grusch answered quickly and directly that what he called *biologics* were, in fact, recovered with some of the craft.

Mace then followed Grusch's reply with the million-dollar question that the world was waiting to hear. Were the recovered bodies human or nonhuman?

Once again Grusch answered quickly, and in a direct and matter-of-fact manner. He stated that the bodies were, in fact, nonhuman.

*And there it was. Just like that, the answer to one of the greatest mysteries of our existence was there for all to hear.*

Facing a penalty of perjury if he answered in a way that was anything other than truthful, the combat veteran and former intelligence officer at the National Geospatial-Intelligence Agency, with one of the highest possible security clearances in the nation, stated to the committee and everyone watching the televised proceedings that not only are we not alone in the universe, but we've had contact with nonhuman species from other worlds for decades.

With David Grusch's testimony, he gave credibility to the idea that advanced beings are part of our reality. He also gave credibility that a higher intelligence may have played a role in our beginnings and in tweaking our evolutionary process along the way.

The credibility of Grusch's testimony supports, at least to some extent, the documented cases of modern humans being abducted, and the warnings they have received and shared regarding our use of technology.

## SOMETIMES YOU DON'T KNOW WHAT YOU HAVE UNTIL IT'S GONE

There is an unsourced adage that says *You don't know what you've got until it's gone.* Another version of this idea is: *You don't know what you have until you lose it, and once you lose it, you can never get it back.* Without a doubt, this idea certainly is true when it comes to us, our biology, and our DNA. What the transhumanist thinking seems to have missed is that we are the product of an ancient genetic formula that is fine-tuned and honed to give us all that we need to thrive in the environment that we live in.

Our immune system, for example, has 200,000 years of history and the lived experience of 10,000 generations that has refined our ability to recognize and neutralize microbes that are not good for us. Our body knows precisely which antibodies to activate, in what order to activate them, and what priority level is needed to apply them for us to stay healthy and go about our daily routines.

Our immune functions, our cellular "software," enables multiple levels of response to anything that our body identifies as foreign to our system. The first-responder cells in our blood, our *innate* immune response, act broadly and quickly, within mere minutes after detecting the threat of a virus, for example. A second class of cells, our *adaptive* immune response, then act more slowly and more precisely to target the invader that has been identified. These preprogrammed T cells have the capacity to recognize various types of viruses, such as coronaviruses, and will begin to multiply rapidly to overwhelm the virus with the best defenses the body can muster at the time. They also activate another class of cells, the B cells, which produce antibodies specific to the viral threat.

After the initial threat has been neutralized, a smaller number of T and B cells remain alert and active, ready to respond to the same or a similar invader at some point in the future. I'm using the complex yet elegant simplicity of the human immune system as just one example of the blood, organ, hormonal, and emotional formula that has been honed to elegant perfection in our bodies over the course of the past 2,000 centuries.

The transhumanist engineers designing tech today believe they know enough now to hijack this advanced system and program the body's own cellular machinery for alternative purposes. And their idea that they can do so without creating negative implications for the rest of the body is naïve at best.

Examples of what I mean here are documented in recent peer-reviewed studies, such as the paper published in the journal *Proceedings of the National Academy of Sciences of the United States of America* in May 2021, showing that modifications to our genome produced by new mRNA platforms, now being used to deliver a number of vaccines, can program our cells to continue making the very toxins—such as spike proteins—that we're asking our bodies to defend us from.[28]

**PURE HUMAN TRUTH 58:** The inherent danger of choosing the transhumanistic path is that the changes engineered into the human genome, once made, cannot be reversed.

The most frightening part of this kind of cellular manipulation is that the genetic instructions, once introduced into the body, can become uploaded (reverse transcribed) into the nucleus of our cells to become a permanent part of our genetic code. In other words, the instructions can tell the body to remain in hyper defense mode, producing the "side effects" that mimic the virus and the infection, for an unknown period. This example is just one instance illustrating the danger of transhumanistic practices.

In my opinion, there are some things in nature that we simply should not disturb. The human genetic code is one of those things. Tweaking this sacred code that was given to us long ago, which gives us empathy, sympathy, compassion, emotion, and the ability to self-regulate our biology, should be a proverbial line in the sand not to be crossed when it comes to the transhumanist agenda. This code was sealed with the signature of the intelligence that gave it to us.

## WHAT'S LOST IS LOST

A species is defined by the specifics of its DNA. If the genetic code for that species is changed and the alterations are significant enough, the species itself becomes a new species. There is a finely tuned genetic balance that allows for any species of life, including humans, to perpetuate itself. When that balance is upset, the continuation of the species is no longer possible. This is precisely how you lose a species. And this is the threat that we now face as the presence of the technology to modify and edit our genome threatens us, our children, and the future of our species.

There is a clear and present danger of unregulated transhumanism in an effort to "perfect" the human experience. As we saw earlier in the chapter, the adage states, *You never know what you have until you lose it, and once you lose it, you can never get it back.* Once we give our humanness away to the technology, we can never get back what we've surrendered. There is no off switch that would allow us to say, "Oops, we don't like the species we've become, let's go back to where we were before we embarked upon the transhumanistic path."

Once we sacrifice our humanness for the logic, speed, and efficiency of computer chips and advanced AI, we will no longer be purely human. I'm not sure as to precisely what we will call ourselves then, but the term will not be *Homo sapiens*.

## PROGRESS OR SURRENDER?

The question posed by the title of this chapter is this: Whose idea of progress are we following? We're at the crossroads of a path that was inevitable once we successfully mapped the DNA of our species. We always knew we would have to make a choice about when, and to what degree, we would intervene in the basic building blocks of human life.

Our technology has now advanced faster than the morals and ethics of how we proceed from here. It's as if we're at the starting line of a race, equipped with everything we could possibly need to complete the race, and we're waiting for the signal to "*Start!*" What's now lacking are the guidelines that tell us how far we should go and how fast we need to go to get there.

But just as it is with any new endeavor, the race begins long before the racers reach the starting line. There are months and years of training and conditioning that prepare them for success. In many ways, the same thing applies to our race toward transhumanism. We've achieved some of the technical milestones to get us started. But, as a society, we've never really decided when or how to apply them.

And now we're in it. We're in the race, and the only way out of it is to go through it.

Fortunately, we have the following three facts to guide us along the way:

- **Fact 1.** We now recognize that without a course correction in the way we use technology, we may be the last generation of *Homo sapiens*.
- **Fact 2.** We've been cautioned by future humans and advanced species regarding our next steps in transhumanism and human-hybrid technology.
- **Fact 3.** We are now aware of our divinity, and that we are more than lucky evolutionary biology. These are all part of the good news.

The information deck is now stacked in our favor, giving us an evolutionary edge to celebrate our humanness and triumph against our transhumanistic challenge. We're now aware that we have a choice as to whose idea of progress we follow. Now it's all up to us. All we need to do is to choose.

**PURE HUMAN TRUTH 59:** The new discoveries of human biology and its divinity now give us everything we need to preserve our evolutionary heritage and make healthy choices that honor our humanness.

Will we become the hybrid human-tech species that visionaries have foreseen for our future? Or will we cherish our humanness enough to protect and preserve what we've been given? We'll know the answer to these questions soon enough. We'll know by the legacy of our humanness that we leave to our children, and they to theirs.

CHAPTER SEVEN

# Deprogramming

## Breaking the God-Spell of Technology

Power is in tearing human minds to pieces and putting them together again in new shapes of your own choosing.

— GEORGE ORWELL (1903–1950),
ENGLISH NOVELIST, POET, ESSAYIST, JOURNALIST, AND CRITIC

To solve a problem, we must be honest about the problem.

This statement is true when it comes to our most intimate relationships, it's true in the boardrooms of corporations and in the classrooms of universities, and it's true when it comes to the ancient and ongoing assault on our humanness and the battle for our divinity.

Without getting lost in the myriad "rabbit holes" of controversy that can distract us from our focus, it's useful to recognize the tactics that we're being subjected to in our quest for the truths that will free us to make healthy choices for our own bodies and in our lives.

I've written this chapter in an effort to inform you of how we're being led to accept disempowering information, as well as addictive technology, in every area of our lives through the high-tech marketing of false narratives designed to skew our beliefs.

Below, I'll identify three powerful methods of indoctrination that we're all being subjected to, explain how to recognize them whenever and wherever they appear, and teach you how to deprogram your mind from these information "spells" that have a grip on you and our society.

## IT'S NO LONGER A SECRET

The tactics are no longer hidden. There is an obvious effort underway to remake our lives, our society, and the world. At the time of this writing, the momentum to remake the world is accelerating. There's an increased sense of urgency on behalf of those pressing for the sweeping changes to remake our society and our lives. It's as if we're on a schedule—a countdown—and the target date appears to be the year 2030.

It's by 2030 that the United Nations' 17 goals for "sustainable" development, as described in Chapter Six, are to be achieved. It's also 2030 when the leadership of the World Economic Forum intend for their goals of the Great Reset to be in place. And, as WEF founder and executive chairman Klaus Schwab emphasized in his 2020 Davos keynote address, participation in the social engineering and digital and economic reforms required to accomplish these goals cannot be optional. "Everyone must participate," he said passionately from the stage.[1]

Schwab left little doubt in the minds of the attendees as to who "everyone" is, and the degree of cooperation needed to achieve the ambitious goals. He clarified, "To achieve a better outcome, the world must act jointly and swiftly to revamp all aspects of our societies and economies, from education to social contracts and working conditions. Every country, from the United States to China, must participate, and every industry, from oil and gas to tech, must be transformed."[2]

It's because of the all-encompassing scope of the targeted transformation and the methods of marketing and media that are being used to achieve the ambitions of these two organizations that I'm speaking to these topics here. Previous chapters in this book describe how the goals now go beyond a transformation of the world *around us*. The stated goals to fully usher in the Great Reset and the UN's 2030 agenda are directly targeting the world *within us*.

The goals are based upon diminishing our humanness in exchange for digital supremacy and control of all life. To the degree that we accept technology into our bodies that replaces our natural systems and functions, we veil the divine attributes that we cherish as a species. These include our freedom to think critically, to create, to innovate, and to live vital and healthy lives.

As we've heard publicly from the Davos forums, to accomplish the goals targeted by the WEF, it is we humans who must be modified and digitally enhanced to fit into the world they envision. This deviates radically from visions of the past, where it was the world that was modified to accommodate the human condition.

In times past, for example, it was the way energy was produced that was improved to accommodate society's growing demand for electricity. It was the production of food staples like corn, wheat, and soybeans that was stepped up to meet the demands of a growing human population.

I'm not saying that these methods were necessarily good things, or the best ways to accomplish the goals. I'm offering these examples to make the point that, until now, we've been altering the world around us to meet the growing demands of our societies, industries, and economies.

With the thinking of the Great Reset, all of that has changed. The idea that we must change—that the way we process information and even our bodies must change—is now the meme that is being reinforced and normalized through social media that is programming the minds of the youngest members of our society.

## THE COST OF SCIENTIFIC ARROGANCE

There are two obvious points of arrogance reflected in the Davos vision of the world.

1. We are inherently flawed as a species and our natural humanness needs fixing.
2. Present-day scientists have the knowledge and technology to fix the supposed human "flaws."

History shows that both of these are dangerous assumptions. And experience shows us that both assumptions are wrong.

While our technology has definitely advanced to the point that we can edit human genes, we appear to know only enough to get ourselves into trouble. For example, genetic scientists definitely know enough to modify gene coding, as evidenced in their cloning experiments on sheep and cows, but the outcomes of their experiments demonstrate that they do not yet know how to do the cloning process 100 percent correctly. The bodies of the cows and sheep that resulted began to break down at approximately 50 percent of their species' natural lifespans. The animals died prematurely from rare, unusual, and painful diseases.

A recent example of the rush to incorporate incomplete scientific knowledge into the treatment of human beings is the 2020 rollout of new mRNA-based gene therapies without the extensive testing that would typically be required before the use of new technologies. The sense of urgency regarding the Covid-19 pandemic became the justification for the briefer-than-usual test times. Numerous reports now reveal that these therapies are implicated in a host of dangerous side effects, including degraded heart function,[3] degraded immune response,[4] higher incidences of blood clotting,[5] and the potential to alter the genome of the recipient (reverse transcription observed in the laboratory),[6] as well as additional symptoms.

These well-documented real-life examples demonstrate that science does not have all the answers. The current models of our humanness do not fully account for the time-tested, delicate, and finely tuned sensitivities of our bodies. Clearly the decision-makers we rely upon to guide us cannot always be trusted to make the best choices when it comes to modifying the vital functions of human life.

This is important to admit, because the media campaigns initiated by the Davos Foundation promote transhumanist modifications that would require us to "enhance" our biology so that we can fit into the Great Reset. The brokers of this plan envision us becoming part of a large-scale digital landscape with our lives enmeshed—literally digitized—into a massive global database of information. The modifications they want for us include all aspects of our lives, from the food that we choose for ourselves and our families to our daily travel habits, how we spend our money and what we spend it on, and to what degree we are able to share opinions and ideas.

The ultimate vision of the transhumanists can only be accomplished if we begin using digital interfaces placed *within* our bodies. It's the information that is detected from us, then gathered and shared by implanted digital devices, that will be fed into the AI-driven systems that Schwab references when he describes the Fourth Industrial Revolution.

**PURE HUMAN TRUTH 60:** The vision of the Great Reset requires us to accept technology into our lives and into our bodies in order that we may become integrated into a large-scale digital landscape.

If this plan sounds like an excerpt from George Orwell's prophetic book *1984*, it's not surprising. The following examples of using AI to identify, locate, and track suspected criminals are about as *1984*-ish as we can get. Even though agencies of our society are still in the early stages of using such technology, the following examples give us insights into the problems that are possible. They show how it can go terribly wrong.

## AI IS FAR FROM PERFECT

While the use of machine intelligence definitely holds promise for applications such as the rapid scouring of historic records to find cancer-healing plants, the technology has not developed to the point that we want to relegate our life-or-death decisions to it. When the AI known as ChaosGPT, for example, was asked, "What is the greatest threat to planet Earth today?" the April 2023 reply was not what the questioners expected.

In a now-famous exchange viewed by millions on the social platform X, the answer was honest, direct, and came without hesitation. "Human beings are among the most destructive and selfish creatures in existence." In the original version of the post—before it was updated in July 2024—the bot went on to state, "There is no doubt that we must eliminate them before they cause more harm to our planet."[7]

Obviously, the algorithm driving this particular synthetic intelligence needs some tweaking, and ChaosGPT was shut down immediately. And while the reasoning of this AI and its expressed intent was largely recreational, there are applications where AI is already being used in real-life situations, with real-life consequences.

In 2020, Robert Williams was arrested in front of his wife and children in Detroit, Michigan, while he was at his home. The alleged crime that he was accused of committing was stealing watches from a local department store.[8] But the reason that the arrest made headlines is because it was not based upon another person witnessing the crime and catching him in the act. Rather, it was a high-tech arrest based upon the use of new facial-recognition software revealing what police believed was a match between Williams's face and that of a known criminal.

The criminal's face had been scanned previously and was being held in a digital database along with the images of thousands of faces of other suspects. Williams's face had been digitally captured, presumably from surveillance cameras within the store, and later processed using the facial-recognition software. When the digital representation of Williams's face was compared to that of the suspects in the database, the software mistakenly "believed" it had found a match between Williams and a known criminal with a history of theft.

Fortunately, this story has a relatively happy ending, as Williams was released approximately 30 hours later when the mistake was discovered. However, at the time of this writing, his young daughter continues to be emotionally impacted by the trauma of seeing her father confronted, handcuffed, and taken from the safety of their family home by police. During congressional hearings on the use of AI in criminal investigations, Robert Williams testified that he and his wife are considering professional therapy to help their daughter to heal the fear and anxiety that she's still experiencing in the aftermath of this ordeal.

If this case was the only time this kind of "glitch" had occurred, it could possibly be written off as the expected and acceptable price that citizens must pay for developing new technology to improve our evolving society. But it's not.

Two years earlier, in 2018, the London Metropolitan Police conducted a test to determine the accuracy of another similar facial-recognition system using a database of people suspected of criminal activity.[9] The results were telling. Only two of the 104 faces sampled by the software proved to be a correct match. This number represents an error rate of 98.1 percent—or conversely, a success rate of only 1.9 percent. Clearly, these are not the statistics of success that we want to base our freedom and our lives upon.

In the Detroit tests described previously, there were at least two additional cases of false accusations and arrests of individuals made after using the same software, under similar circumstances. Fortunately, the men wrongly accused in those incidences, like Williams, were later absolved of the crimes they were accused of committing and released.[10]

However, as in the case of Robert Williams, the two men have continued to endure the social consequences of false accusations in ways that include ongoing family trauma and stigma in friendships and the workplace. And this is the concern.

If this kind of mistake can be made today, using the most basic physical features of facial recognition to match a suspect with surveillance photos, what are the chances of making mistakes with even more dire consequences in the next level of surveillance that uses AI to predict criminal activity, including when and where a crime will happen?

## THE DANGER OF USING AI TO PREDICT CRIME BEFORE IT HAPPENS

The vision for the future of AI is now expanding the role of predictive programming to include the likelihood that a crime will happen, before the crime actually occurs. A 2022 paper published in the peer-reviewed journal *Nature Human Behaviour* describes how researchers at the University of Chicago have successfully predicted the event of a crime to an accuracy of within 1,000 feet of the location where it actually occurred, and have done so within one week of the time the crime actually happened.[11] The scientists who generated this prediction used algorithms based upon historical data gathered over a period of five years.

One of the lead researchers, Ishanu Chattopadhyay, compared the way the predictive AI algorithm works to weather forecasting, saying, "We discover patterns in event logs, and apply these patterns to calculate risk of an event in future at specific locations. It is like if you see dark clouds, you conclude that it is going to rain soon; just here the patterns (the dark clouds) are much more subtle and harder to recognize and reason with."[12]

The hope of predicting this kind of activity is that the patterns can be recognized and used as a guideline when it comes to allocating budgets, resources, and staff in an effort to make crime-prone areas of large cities safer.

## NOT IF, BUT WHEN?

At the time of this writing, the University of Chicago AI models are *not yet* being used to actually predict if or when specific people will commit an illegal act. Chattopadhyay clarifies this, stating, "We do not focus on predicting individual behavior, and do not suggest that anyone be charged with a crime that they didn't commit, or be incarcerated for that. Our model learns from, and then predicts, event patterns in the urban space . . . It does not indicate who is going to be the victim or the perpetrator."[13] I placed the words *not yet* in italics at the beginning of this paragraph for an obvious reason. That reason is human nature.

History shows that once we put the time, energy, and resources into developing a new technology, it's rare that the fruit of the effort is not used at some point. Once we split the atom and learned how to harness the energy this releases, for example, it was only a matter of time, and a brief time at that, before that energy was used to build a highly destructive weapon near the end of World War II.

After the techniques of cloning were demonstrated upon Dolly the Sheep, it was only a matter of time before scientists applied them in an attempt to modify other embryos, such as those of domestic cows. Now that we have the AI that can predict criminal activity with 90 percent accuracy within days of a crime, the proverbial genie will never go back into the bottle. It's only a matter of time before this technology is used to attempt to predict the likelihood of specific individuals performing criminal acts.

In large part, the data to make such a prediction already exists. It has been gathered from sources that already exist and are already in use. These sources include big databases created by the software giants that we all depend upon for our daily computer use, the social media platforms we visit and post on, Internet records of our browsing and purchasing habits, and from public records that are digitally archived. The additional data resulting from the proposed medical applications of transhumanist technology will further refine the access authorities have to the more intimate realms of our lives.

For example, one of the pieces of medical equipment proposed as part of the Great Reset is a cluster of nanosensors placed into the body either via injection, inhalation through airborne particles (sub-millimeter-scale smart dust currently used in meteorological applications), or administered orally. The nanosensors would function like a more sophisticated version of the wearable-tech devices that some people use today to capture data on their athletic performance and others use to monitor the quality of their sleep—such as wristbands made by Garmin, Fitbit, Apple, and Google. The continuous stream of biometric data emitted from nanosensors, such as levels of blood pressure, heart rate, insulin, and stress, we're told, will ensure that we address our health challenges promptly by quickly identifying problems at the source, right when they occur.

It's these same biometric measures, however, that may also be cross-referenced in the digital landscape to the previously identified psychological patterns of individuals, such as anger, depression, and violent family history, to create the profile of a candidate for potentially dangerous behavior.

## AI DOES NOT ALLOW FOR FREE WILL

It's precisely the concerns of data abuse that have been raised by Patricia Kosseim, former general counsel of the Office of the Privacy Commissioner of Canada. In a 2022 paper, "Privacy and Humanity on the Brink," Kosseim identified the potentials of AI to skew public opinion and policies and the thinking of vulnerable young people.

In the paper, she asks the question that must be asked: "Will algorithms and neurosensors designed to predict who is the most likely to commit (or recommit) crimes based on certain sociodemographic factors create a self-fulfilling prophecy, robbing us of our freedom to defy whatever odds may be against us to become the person we want to be?"[14]

When many people hear of the use of AI to predict crimes and criminal behavior, Steven Spielberg's 2002 film *Minority Report* comes to mind. In this sci-fi thriller, based upon Philip K. Dick's 1956 novella *The Minority Report*, intuitively gifted humans are used to sense impending behavior. They predict the crimes that people will commit, and do so before the crimes actually occur.

The main difference between Dick's vision as depicted in the movie and the examples in this chapter is that crime prediction today is being done by computers and AI rather than gifted humans.

In either scenario, however, the outcome is the same. And as Kosseim states clearly in her article, the use of these predictive technologies robs us of our human agency. The prediction algorithms do not allow for the power of our free will to direct our choices, the healing of past trauma to transcend the predictive markers, and our divine agency that allows our healed perspectives to become manifest in our lives. In other words, just because the deck is stacked against us, and just because the markers for social violence may be present within us, doesn't mean that those potentials will necessarily manifest as violence. It doesn't mean that we will necessarily commit a crime.

The same principle applies when we fill out a family medical history checklist in preparation for our annual physical exam. Just because our parents had many—or all—of the illnesses and conditions identified in the list on the clipboard page in front of us, that doesn't mean we will inevitably experience the same illnesses and conditions. Such lists don't allow for epigenetic factors, such as our choice of lifestyle, stress-management techniques, and emotional-healing activities, to offset the family genetic history that is often used in predicting our health. Our ability to express qualities of our divinity, such as love and forgiveness, is a factor that is not accounted for, at present, in the AI algorithms and predictive models.

## THE OBSTACLE

There is only one "problem" that stands between the world of today and the world of the digital future envisioned by the policymakers at the WEF and UN. That problem is you. And it's me. It's our individual lives and collective communities that reflect the divine qualities of our humanness. The very joyous, positive, innovative, self-reliant, and creative qualities that you and I cherish within ourselves and aspire to when we see them expressed by others have now become the roadblock to achieving the integrated data matrix that is the goal of the Great Reset.

> **PURE HUMAN TRUTH 61:** The power, imagination, and freedom of human divinity are seen as obstacles to those who are attempting to achieve the goals of the Great Reset.

The worldly expressions of our divine nature are also expressions of the sacred and even deeper values of freedom: freedom of thought, freedom of sharing ideas, freedom of imagination and creativity, freedom of self-reliance, and the freedom to share with others what inspires us in our lives. Freedoms such as these are so common in the modern world that younger generations often assume we've always had them, they've always existed in the world, and they cannot be taken from us.

Wars have been fought, suffering has been endured, and lives have been lost to ensure the continuation of the liberties we enjoy today, yet the freedoms we celebrate are no longer emphasized and revered in the classrooms of many public schools in America. The transhumanists want us to believe that surrendering these freedoms is the price we must pay to enjoy the benefits of the emerging world of digitized lives managed by smart systems.

In exchange for the sacrifice of our freedoms, we're told that we can expect a mediocre life of homogenous existence. No one will be too poor. No one will be too rich. Everyone will be given the basic necessities of life by the powers that be, which will track, regulate, and control them through the digital interfaces that have become the Internet of all things. "We'll own nothing and we'll be happy" is the slogan that is proudly recited in the now-viral WEF promotional video that has been created to prepare us for this grand restructuring of society.[15]

The idea of the WEF video that has been circulating online is that although you and I will own nothing, we will have everything we need. We'll be relieved of the stress and responsibility of ownership of our worldly goods. Instead, we'll be living in a use-on-demand, rent-as-needed, sharing-based society. When we need furniture or a car, for example, we'll rent only what we need, keep it just for the time that we need it, and it will be delivered to our door by an automated drone or smart driverless van. We'll be carefree and happy as we move from a consumer society to a "kinder" sharing society.

There's an unspoken catch here. While it is absolutely true that, under the system proposed, "we" will own nothing, it's also absolutely true that someone else will own everything. Someone else will own our homes. Someone else will own our cars, our computers, and our smartphones. And whoever that "someone" is will be wealthy and powerful beyond the kind of wealth and power we can imagine today.

This model is the very expression of the kind of elitism and serfdom that George Orwell envisioned in his fictional warnings about what could become of the world. While the year in which he foresaw this dystopian vision was 1984, the specifics of his vision are either in place now or within sight on the near horizon. He was only off by a few years.

## TRADING DIVINITY FOR CONTROL

A close examination of the policies being proposed to implement the Great Reset and the UN's Sustainable Development Goals reveals that they are in direct opposition to the basic principles of divinity, freedom of self-determination, and personal expression. We are spiritual beings, expressing our spiritual energy through the daily actions that we engage in to keep ourselves alive and healthy and keep the engine of the world's economies running

In this context, the word *economy* is not so much about money itself. Rather, it is about people exchanging the information, goods, and services that they need and use in their lives. The spiritual energy that we place into producing information, goods, and services is literally a divine economy that is based in our creativity, imagination, innovation, and freedoms of choice and self-expression.

In contrast, the world envisioned by the Great Reset is based upon a global network of control exercised through all-inclusive surveillance, all-inclusive tracking, all-inclusive monitoring, and all-inclusive restrictions.

Clearly this vision is for society to shift from one of distributed wealth, innovation, production, and consumption to one of centralized wealth and regulated consumption, with innovation being reserved for a handful of people who believe they know what's best for everyone else and for the world. The methods that are being used to accomplish this unprecedented magnitude of social and spiritual engineering are both insidious and brilliant at the same time. And they lead to what may be the greatest irony of all.

The irony is that these changes are not being imposed *upon us* by an invisible outside force acting covertly. Rather, the changes are coming *from us*. We are the ones inviting the changes and welcoming them into our society and our lives. The insidious part is that, with the exception of a few astute and well-informed individuals, most people don't even know what they're asking for. All they know is that life has become difficult and they've been promised an easier way of life if they accept the changes of control in the world, gadgets into their bodies, and the loss of their freedoms.

**PURE HUMAN TRUTH 62:** Under the pretext of providing us an easier life, we are being led to ask for the very changes in society and our lives that are destroying the fabric of society and our lives.

In recent decades, for example, young people in the United States have been taught to follow the philosophies that their teachers bring into the classroom, and to do so without question. This style of teaching contrasts starkly with the education of previous generations who were taught the art of critical thinking so that they could validate the accuracy of the history, politics, and technology they were being asked to embrace in their lives. The lack of discernment produced by the poor quality of public education has led young adults to believe it's okay to enlist gadgets to replace their own creative processes that they've used in the past.

AI interfaces, for example, are now being used by musicians and songwriters to write the lyrics and music for songs that are competing against actual human compositions for awards, including prestigious Grammy Awards. They're being used by artists to enhance, or even replace, images that were once created exclusively by the brushstrokes of a human hand. A report released in late 2023 reported that approximately one half of the school students surveyed said they have used AI to accomplish their assignments, while 62 percent of those students said their teachers did not approve the use of AI in the classroom.[16]

In the same way that younger generations are asking for the very technological devices that are robbing them of their freedoms and self-sovereignty, they are eagerly embracing gadgets that steal away their humanness and their divinity. The sad part is that they don't realize it. The narrative that we humans need something outside of us to fix us, and the cultlike thinking that idolizes technology as our savior, is part of a deeper process, and an example of a new form of social warfare, that has swept through the entire world in recent years.

This war, however, isn't between armored tanks engaged on a desert battlefield in a distant land. It's a war that's raging in the most intimate theater of battle that we humans can experience. This war is happening right now, and it's playing out in the theater of our minds.

## FIFTH-GENERATION WARFARE

We are all casualties of a new style of warfare that's playing out in our minds. The battles have been raging for so long, and their existence has become so normalized, that we typically accept them as a given in our lives without a second thought. It's only recently, however, with the advent of the Internet and advanced social media algorithms that this kind of psychological warfare has successfully swayed public opinion in terms of when we go to war, how we conduct politics, and how technology is being accepted into our bodies. The name given to this kind of warfare is *fifth-generation warfare*, or more commonly Five Gen Warfare (5GW). The 5GW occurs without a shot being fired from a conventional weapon. Instead, volleys of misinformation, disinformation, and flat-out lies compete for the prize of our thoughts and what we believe to be true in our lives.

In his 2010 book, *The Handbook of 5GW*, historian Daniel Abbott describes this kind of warfare as a war of "information and perception."[17] Its value is how it triggers people to destroy themselves from within. Once people are exposed to psychological warfare, they destroy their own societies, their own nations, their own communities, families, and even their own bodies from within. They do so without the need for an oppressive army to impose destruction upon them. They do so through the choices they make and the policies they demand, which are based on the information they've been led to believe is true.

**PURE HUMAN TRUTH 63:** Fifth-generation warfare plays out in the battlefield of our minds before there is ever a kinetic engagement between people.

By creating confusion, uncertainty, doubt, and fear of one another, of government policies, of the judicial system, and of leaders, it's the population itself that shatters the psychological bonds that have long given them common ground, comfort, and safety in the past. It's the destruction of these common bonds that make people vulnerable to alternative visions that remake their lives, and ultimately our world. This kind of warfare is so prevalent now that it plays a role in each of our lives each and every day.

Every day, we interact with friends, family, and co-workers, sharing ideas with them that are based upon what we believe to be true. The ideas that we base our opinions upon, more often than not, come from the social media feeds of cable and network "news" programs—all of which are distorted and skewed by algorithms put in place by owners of these channels who allow or deny us information in order to support a specific narrative.

To some degree, we are all casualties of 5GW. The brilliance of this style of warfare is that most people don't even realize that a battle is underway, or that they're a part of it. The concerted effort being made by political enemies to break the social bonds that have made the United States a strong nation in the past is a perfect example of what I mean here. I call their efforts to destroy our self-image the *Dangerous Game*.

As one definition of a game is to "manipulate a situation, typically in a way that is unfair or unscrupulous," and because we are all inundated, these days, with information designed to skew the choices we make, in a very real sense, we are all playing the Dangerous Game.[18]

## THE DANGEROUS GAME

The Dangerous Game is playing out in front of our eyes, and it's happening right now. The game is dark. The game is hurtful. And because it's based in fear and hate, it's ravaging our friendships, marriages, families, and communities and disrupting our trust in our own bodies.

The current version of the Dangerous Game is a large-scale social engineering project that is separating us from our traditional values and beliefs and diminishing our personal sense of security. Just as in any game, in the Dangerous Game there are winners, there are losers, and there is a goal, and the strategies of the game can be tweaked as necessary to produce specific outcomes.

The goal of the Dangerous Game is simple: it is designed to pit people against people, to divide us from one another, to break the bonds of our social coherence, and to destroy the unity of our families and communities. The game ultimately destroys our own sense of uniqueness, our worth, and our specialness as humans. When we doubt and lose trust in ourselves and our sense of divinity, we become lost.

The purpose of the Dangerous Game is clear: to cause us to doubt and lose trust so that we feel lost. In our disarray, it becomes impossible for us to work together for our common good and to solve our problems together. This is where the Dangerous Game comes in. When we feel like we can no longer solve the problems in our own lives, then we are willing for someone else to step in to make choices and implement solutions for us.

The key here is that in our inability to act consciously and with confidence, new solutions are *mandated* by another authority. Because of our sense of powerlessness, things that we would never agree to or accept under normal circumstances can be imposed upon us by those who have power over us and happen in ways that we have not agreed to, or chosen.

**PURE HUMAN TRUTH 64:** The goal of the Dangerous Game is to be led to diminish one another—to destroy the trust and social bonds that hold together our most intimate relationships, our families, and our societies until we are led to question and doubt our own value and humanness.

The Dangerous Game has been played for centuries. Through the skillful use of misinformation, disinformation, and lies, our darkest fears and our most primal instincts for survival are skillfully ignited and manipulated into the mistrust and hate that turn us against one another.

The game begins when a false narrative is read on social media, heard from a friend, or viewed on network television under the guise of "news." If the narrative is believed and accepted as true, then relationships may be tested and social division begins.

## WHEN OUR DIFFERENCES BECOME WEAPONS

At first, the break in our social alliances and self-image can appear as a subtle fracture in our ability to trust other people, which also diminishes our self-confidence. While these doubts have always existed, today the diversity of race, financial status, and education that we once celebrated as strengths of our way of life are now emphasized as inequalities to be criticized, ridiculed, and feared.

In other words, the very attributes that we cherish in our humanness are weaponized to appear as the qualities that make us obsolete.

**PURE HUMAN TRUTH 65:** Through the Dangerous Game, differences in race, religion, culture, and diversity that we have celebrated as strengths in the past are weaponized, leaving us divided and vulnerable to the ideas and agendas of others.

Once the Dangerous Game is underway, people who have a viewpoint that does not conform to the accepted beliefs of the majority of the population around them may find themselves being shunned, shamed, demonized, and rejected by the people they know, including friends, relatives, neighbors, and co-workers. Their reputations are often attacked and attempts to ruin their businesses are made under the pretense of leveling the playing field through "street justice." In the United States today, we call this *canceling* a person.

## WE'VE SEEN THIS BEFORE

The Dangerous Game is nothing new.

We've seen it in the past. We saw it in Europe in the early to mid-20th century as people were *taught* to hate their neighbors with Hebrew heritage. As a result of hatred, an estimated 6 million Jews were systematically exterminated, along with approximately 5.7 million Soviet civilians, 2 million Poles, 4.5 million Serbs, 300,000 Romani, 10,000 people in same-sex relationships, and over 3,000 Jehovah's Witnesses, among others.[19]

We saw it in 1994 in Africa as the Rwandan Hutus were *taught* to hate the Tutsis in class warfare that led to civil war and a genocide that lasted 100 days, killed between half a million and a million people, and destroyed approximately 70 percent of the Tutsi population.[20]

We're seeing it today in the attempt to separate and divide us from one another by using distorted narratives and disinformation that *teaches* us to view one another with distrust and to hate one another—for example, Christians against Muslims, Muslims against Jews, Whites against Hispanics, and Blacks against Whites, the rich against the poor, and men against women. We're seeing this very tactic used to create confusion between the sexes and the perception of what it means to be male and female.

When it comes to the transhumanist agenda of the Great Reset, the tactics that have proven successful in each of the previous instances are also proving successful in casting doubt on the value of our humanness and what the lineage of our species represents. We're seeing the Dangerous Story dividing people against people, and individuals against themselves, again for one reason—because it works.

## THE TACTICS

The tactics that make such confusion, uncertainty, and division possible are timeless. They're taken right out of history's playbook for dividing people and destroying societies. Eighteenth-century philosopher Immanuel Kant, for example, simplified the long-recognized philosophy of creating division among the population that others want to control. In his 1795 book, *Perpetual Peace: A Philosophical Essay,* he described the specifics of the technique that has since become known as "divide and rule," or more commonly as "divide and conquer."[21] In his Appendix 1, "On the Disagreement Between Morals and Politics with Reference to Perpetual Peace," Kant states key themes of this technique, which include:

1. Creating divisions to prevent alliances that can challenge the ruling class
2. Fostering distrust within the local population

More recently Kant's principles have been incorporated into the philosophy of community organization, illustrated in the 1971 book *Rules for Radicals* by activist and author Saul Alinsky. In his book, Alinsky identifies 10 rules for effective social engineering, with one chapter devoted to each of the rules. In summarizing one of the 10 rules, for example, Alinsky condenses Immanuel Kant's philosophy for divide and conquer into four simple steps, stating that the key is to "pick the target, freeze it, personalize it, and polarize it."[22] In not-so-subtle ways, this is precisely the template for what we see playing out in the world today.

Clearly, there are social injustices in the world that need our attention and care. Real and lasting solutions, however, cannot happen if we attempt to solve them as divided families, polarized communities, and as a species that believes we cannot be redeemed from the horrors and atrocities that have been part of our past. The Dangerous Game attempts to reduce the power of the common bonds that have connected us in the past, while weaponizing the differences or "flaws" that burden us in the present.

## DEPROGRAMMING: BREAKING FREE FROM THE DANGEROUS GAME

The only way to triumph over the dehumanizing goals of the Dangerous Game is for us to choose ways of thinking that transcend its goals and tactics. We must *deprogram* ourselves from the marketing tactics and our technological addictions, just the way that any other addiction must be healed. And to do so, we must begin by recognizing what it is that we're up against.

From those who think they're "in the know" with the true inside story of what's happening to themselves and to the world, to those who don't have a clue as to what's happening and couldn't care less—one way or another, we're all in this together.

To end the game means to break free of the godlike spell of technology and social media programming that we have been enticed to accept. This means having the wisdom, courage, and strength to (1) turn away from the false narratives and misinformation that are encouraging us to give our power away, and (2) recognize how we give our power away by accepting advanced technologies, such as invasive social media and synthetic chemicals, into our lives and bodies.

It also means that we must recognize that some of the most popular technologies that have become commonplace in our lives are the very technologies that have been hijacked to shape our perceptions, and thus, engineer our consciousness.

## HISTORY RHYMES

It's been said that while history itself may not repeat, it definitely "rhymes." While we may not see the big events of the past repeating themselves *exactly*, we definitely see the themes of past events playing out in successive generations. The reason for this is that while the world marches forward to grow and change, human nature seems to progress on a slower trajectory. History shows that the same ideas of power and domination are playing out today that have set the course of our past and have done so for generations.

This is true for the use of information to sway public opinion to support policies that otherwise would typically never be accepted, such as engaging in acts of war.

We now know, for example, that the Gulf of Tonkin incident that was used to justify the American escalation of the Vietnam War in 1964 was a "nonincident." It never actually happened. On August 2, 1964, there was a brief confrontation between U.S. naval forces and their Vietnamese counterparts near North Vietnamese territorial waters. Wanting to sway public opinion in support of military action, American intelligence agencies intentionally skewed the narrative of events that followed the encounter to make it appear as though a second confrontation had occurred two days later on August 4.

In an interview with Errol Morris for the 2003 documentary *The Fog of War*, former U.S. Secretary of Defense Robert McNamara, who served 1961 to 1968, acknowledged that there had been no second attack on August 4.[23] Two years later, in 2005, documents were declassified that confirmed McNamara's statement, revealing that the intercepted communications had been misrepresented to appear as though a second attack on U.S. forces had occurred.[24]

Based upon the false narrative that was reported through the limited media sources of the day (they had print, radio, and television journalism, but no Internet), the public believed that American forces had been attacked and that an escalation of the war effort by the government was justified. These perceptions led to the deployment of an additional 100,000 U.S. troops in March 1965, another 100,000 in 1966, and the destruction of countless Vietnamese villages and vast amounts of farmland. When all was said and done, the human cost of the war between 1965 and the war's end in 1975 was the death of approximately 65,000 North Vietnamese civilians and 58,220 U.S. servicemen and servicewomen, as reported by the U.S. Department of Defense a few years later. (Today, the number of civilian lives lost is understood to have been much higher.)

More recently, the false narrative that Iraq's president Saddam Hussein was harboring "weapons of mass destruction" was used to justify the U.S.-led invasion of that country in 2003. As a consequence, the nation's infrastructure of electricity, water, and civil services was destroyed, the Iraqi government was dismantled, and the still-disputed casualty numbers ranged from between 186,000 and 210,000 civilian deaths.

The reason that I share these statistics is that entering each of these wars needed the support of public opinion to move forward. People are generally opposed to war, the killing of other people, and the spending of tax dollars to fund a wartime effort. Without the engineered narratives, the wars could not have happened. Media-based social engineering is not limited to the war efforts of the past. We see precisely the same principles playing out today in arenas beyond war that are closer to home.

Examples of media-based social engineering include: narratives surrounding the use of pharmaceuticals to slow, or even stop, puberty from occurring in adolescents; narratives about the relationship between the races in America; narratives about the relationship between people of different faiths; narratives about climate change and the reasons why it is occurring; narratives about what we need to do to get and stay healthy; and so on.

Each of these issues is important to the people whose lives it touches. Each one impacts our lives and our nation to some degree. And each one needs to be discussed in a way that is honest, truthful, and factual. Unfortunately, important issues that people feel strongly about in different ways are being weaponized through social-media and search-engine digital algorithms that boost the number of posts and articles we see about them aggressively echoing our own opinions, the outlets that host the media, and the personalities and familiar faces who are paid to promote skewed perspectives.

In this way, our consciousness is being shaped. The power of our humanness is being hidden and our perceptions are being manipulated in ways that veil the true power of our divinity.

## HOW TO UNPLUG FROM THE ADDICTION TO SOCIAL MEDIA

To break the technological spell over us is to change the rules—to dream a new dream and create a beautiful, healthy new world together.

When we stop allowing ourselves to be distracted and triggered by the media clickbait that lights up our deepest fears and our greatest anger, and offends our common sense and human sensibilities, the Dangerous Game is over. It only continues as long as we continue to play it.

The threat we are facing right now, together, is bigger than partisan politics. It's more than left-wing or right-wing posturing, more than conservative and progressive philosophies. It's more than the outcome of a single presidential election. These are the outward and obvious distractions that are expressions of the game. In a very real sense, our indoctrination into the Dangerous Game is a form of psychological programming.

The steps that make it possible for us to unplug from these unhealthy programs resemble the same steps that are needed to deprogram from any structured form of indoctrination, or what has been called brainwashing in the past. In the following, I will briefly identify some of these deprogramming steps so that you may be aware that they exist and the role they play in your life, and apply them to yourself if you feel drawn to do so.

Steven Hassen is an American mental health counselor who gained notoriety in the 1970s for developing psychological techniques to help with deprogramming the thinking of people who got caught up in any of the various cults that were popular at the time.[25]

Based upon the experiences that Hassen had with the deprogramming process, he has since developed a novel deprogramming philosophy based upon noncoercive techniques of developing an awareness and context for the indoctrination, which any of us can use when we feel like attempts are being made to sway our thinking.

Following are descriptions of a few of the tactics that are typically used for the indoctrination of individuals into cults. I'm sharing them here because they illustrate direct parallels to the kind of indoctrination that people in America, and in much of the world, have been exposed to in recent years. I'll first identify the tactic, and then suggest possible remedies to deprogram your thinking from the indoctrination.

## Indoctrination Tactic 1. Present the Doctrine as Reality

This tactic attempts to persuade individuals that the information they are receiving is accurate and "the" one and only truth. In doing so, it attempts to invalidate any information that does not agree with or support the perspective embraced by the individual.

One of the ways this indoctrination tactic is applied in our high-tech society is through the formation of social media *information silos.* Personalized information links are created by algorithms of a social media platform so that it can "learn" the kind of information that we are looking for in our online searches. The algorithm will then seek out additional sources of information that agree with the viewpoint we have shown an interest in and topics we have just searched for. Ultimately, this process fills our web pages, e-mail inboxes, and texts with advertisements, book suggestions, and invitations to webinars and seminars that reflect the perspectives we have shown an interest in.

*The key here is that the algorithm will continue to feed us links to information that supports our viewpoint, and do so to the exclusion of any other view or perspective.* The result of seeing so many similar messages, showing up in so many digital channels, is that we believe our viewpoint is so pervasive that everyone else is seeing the same thing we're seeing. We believe that it must be the truth.

At the same time that we're seeing our personalized information silo, our friends and loved ones are experiencing their own information silos. They believe with equal certainty that the information they encounter is the truth. This experience can make it difficult to entertain any viewpoints that may question or contradict the information we've received. It can, and often does, create a separation between us and friends and loved ones who are equally deeply entrenched in their information silos.

## The Remedy for "Present the Doctrine as Reality"

While there is no single answer to remedy the "Present the Doctrine as Reality" tactic, there are a number of solutions. Some people simply unplug from web-based information and social media altogether.

While this can be effective in the short term, it may isolate you from important information that you need to participate in your day-to-day routines.

If you choose to remain engaged online and you want to do so responsibly and objectively, the most effective remedy to this tactic is to seek out alternative and even unconventional sources of information. Multiple points of view is what you're after. Using your sense of discernment and doing so responsibly, you owe it to yourself to explore various news feeds offering various perspectives, knowing that, ultimately, the reality of whatever event you're exploring is probably somewhere in the middle of all that you're seeing.

For example, I followed the horror of the events of October 7, 2023, in Israel while I was speaking at a conference in London. Due to the time difference between Europe and North America, the first news feeds that I saw came from the Middle East, India, and Europe.

Within minutes of the events that had happened, I was seeing full-on the magnitude of the atrocities being perpetrated, in some cases through the eyes of raw and unedited footage. By the time the American news stations began reporting the stories later in the day, however, I could see the differences in how the events of that same day were being presented to U.S. audiences. With few exceptions, American media outlets refused to show the images or to discuss the horrendous details of the unspeakable atrocities that had occurred only hours earlier. The community rules and community guidelines of social media platforms also would not allow the images to be shown uncensored.

The result of these actions was that many Americans were, and continue to be, falsely led to believe that the factual events that occurred on October 7 were not real and that the atrocities never happened.

The repercussions of this manipulation of information continue, and the consequences are playing out on the global stage today.

## Indoctrination Tactic 2. Manipulation through Fear and Guilt

This tactic is particularly insidious, as it plays directly to our instincts for altruism and survival, and then pits those instincts against a false sense of guilt for our perceived role in contributing to the suffering of others. In recent years, this technique has been used to persuade communities and nations to make radical changes in their policies for everything from energy use, urban design, and farming techniques to land use and medical care. Misinformation is "weaponized" to instill guilt and fear and rally people to a cause.

I'll use the phenomenon of our changing climate, and the way that the information about it is being presented to the public, as an example to illustrate how this tactic works. Through the use of information silos, the public is educated in a specific narrative regarding climate change. This perspective is reinforced to the point where people believe they are well informed as to both the cause of climate-based disasters, and the remedy that is part of the agenda the indoctrinators are pursuing.

The sequence of events associated with indoctrination looks something like this. First, our information silo shows us the very real, the very horrendous, and very frightening events, such as torrential rains creating floods, hurricanes, and tornadoes destroying low-lying communities and coastal cities, and uncontrolled wildfires raging through urban communities where the flames destroy people's homes and businesses and disrupt or sometimes even end their lives.

We are naturally shocked by what we see. Our human nature wants to help in some way. We want to alleviate the suffering that we're seeing. We often feel powerless to do anything in light of the sheer magnitude of the disasters.

Until this point, what we're seeing is real. The destruction and loss of life is tangible, visible, and undeniable. What happens next is where the manipulation begins.

Taking advantage of our human sensitivity and building upon our emotional vulnerability after having witnessed suffering, we are immediately led to believe that the horrible events we're witnessing are, first, the direct result of climate change, and second, that this change is happening primarily because of us—our behavior. We're told the suffering is happening because of our choice of fossil fuels as the primary source of energy for our civilization.

While honoring the data available to us, which makes the fact of our changing climate impossible to deny, the problem is that the reality of the change and what it means in our lives has been weaponized. Rather than using our technology to adapt to the rhythms of natural climate change that have been anticipated for decades, the phenomenon is used to divide us as families and communities as described previously.

Contrary to the public narrative promoted in mainstream media and classrooms, the geologic data from diverse sources reveal that Earth's climate is not static. It's a dynamic and complex system that is constantly changing. The changes we're seeing today in global climate and localized weather patterns are (1) part of Earth's rhythms and cycles seen in the past, and (2) that the bulk of the increased $CO_2$ in the atmosphere comes from the outgassing of oceans rather than from the burning of fossil fuels.[26] Without a doubt, $CO_2$ levels are definitely higher today than they have been for recent decades. And without a doubt, today's $CO_2$ levels actually pale in comparison to levels they've achieved in ages past.[27]

It's no coincidence that the periods of geologic history that saw high levels of $CO_2$ also correlate with the times that Earth experienced lush forests and a healthy diversity of life. These demonstrable facts are in direct contradiction to the frightening predictions coming from "experts," misinformed activists, and many of our political leaders today.

Clearly, it is to our benefit as a species to develop clean, green, and sustainable forms of energy that are available to everyone that wants access to the energy. If our leaders were serious about doing so, however, the technologies discovered over the last 70 years would have been developed and made available to replace the fossil fuels that we rely upon today.

These include the use of the element thorium, for example, to power baseload electric power plants as validated during the

Manhattan Project in the mid-20th century; allowing the carburetors developed in the 1970s that achieve 90 miles per gallon of gasoline to be used in the automobile industry; and access to the resonant technology that draws energy from the Planck vacuum of "empty" space to be made available for commercial use.

*Note:* While it is beyond the scope of this book to share the evidence for the healthy impact of higher $CO_2$ levels in detail, I've done so extensively in online video presentations. You can find links to these on my website, GreggBraden.com.

With climate narratives manipulated in the way I've described here, it's not surprising that we find ourselves instilled with sadness, anger, and fear from the narratives that permeate the media today. It's also not surprising that we feel guilt regarding the possibility that we have somehow contributed to the horrors and suffering of what we're seeing.

The most seductive part of this kind of manipulation is how it uses a mix of factual, real-world events to promote guilt and then to justify extreme forms of social engineering as a means to an end. The means detracts from our lives by destroying entire industries and traditional ways of life and living, while the end we are headed toward will enrich a select few who know how to build and invest in technology and financial instruments that align with mandated changes.

The sad irony is that all of this is happening while having zero impact upon the climate extremes that have caused the tragedies to begin with.

It's been said that knowledge is power. But simply having knowledge of a subject is not enough. It's the kind of knowledge that we have access to that is the critical factor in protecting yourself from indoctrination by those who do not have your best interests at heart. If the new information we receive is coming to us through the same media filters that fed us the original distorted information that we accept, the new information will simply reinforce the point of view that we've already accepted.

This is precisely why the battle for our thoughts and beliefs rages on a daily basis in the output of our newspapers and magazines as well as television and radio broadcasts. The magnitude of social engineering that is being attempted based upon the false narratives of climate change is making the availability of objective information even more valuable than ever. The better informed we are, the better equipped we are to make informed and responsible choices.

## The Remedy for "Manipulation through Fear and Guilt"

The simplest remedy for being manipulated by guilt and fear is to check your facts. Fear largely comes from confronting an unknown, and guilt from our sense that we've done something that hurts someone else. Both of these feelings result from what we believe to be true in a situation—what we've read or heard from someone else. This is why it's vital to discover the reality of what we're up against whenever information that comes to us makes us feel guilty and afraid.

This is precisely why the verifying of facts is so key to our deprogramming—or preventing brainwashing in the first place.

Automated "fact-checking" software is probably not a good place to begin. Personal experience has shown me that the online fact-checkers and AI are often under the influence of the same algorithms that created the media silos we are living inside, reporting misinformation to us. Also, as search engines reach into the datasphere of the worldwide web to answer our questions, they are drawing upon the information that has already been skewed and distorted by algorithms to begin with.

While doing so is a bit more tedious, it is still possible to do the research today that makes our due diligence effective. This can be accomplished by:

- Searching for direct research and scientific reports yourself rather than relying upon podcast commentary on those reports or a news anchor's interpretations to explain their meaning to you. While the papers and articles we access in this manner may be more technical reading than you're accustomed to, you'll be reading the direct conclusions of the scientists and researchers themselves—in the intended context—rather than the commentary of an uninformed personality.

- Searching for in-person, phone, or e-mail interviews with experts so you can hear directly from individuals themselves what their most up-to-date views, opinions, policies, and conclusions are. Ask clarifying questions from the audience when you attend lectures and workshops.
- Being open to alternative sources of information beyond typical mainstream news feeds and broadcasts prepared by popular media personalities whose voices you are already accustomed to hearing. This includes reading, listening to, or watching reports made by independent journalists, including some from countries whose policies you do not necessarily agree with. Remember, as you do, that this is an exercise to examine all sides and perspectives of an event objectively, rather than forming your opinion from the narrative that is easiest to access and closest to your opinion.

## Indoctrination Tactic 3. Ensure There's No Way Out

The fear that there's no way out of an extreme and frightening situation is a hybrid tactic of indoctrination. It couples the "fear" aspect of tactic 2 with a sense of hopelessness described here in tactic 3 to make it appear that there is only one conclusion to what we're being presented with. We must conform, submit, or "cave" to the agenda of the powers that are responsible for the indoctrination. This sense of a binary, either-or situation is typically linked to a false sense of urgency that has been created by a false narrative based in partial truths.

The response of leaders during the global pandemic of 2020 and the climate-change scenario we discussed previously are both examples of what I mean here. In both situations, based upon the information that was repeated and widely distributed by trusted authority figures we had relied on in the past, we were presented with scenarios that sounded scary, told these would have dire consequences, and then were told that we don't have time to debate the significance of the emerging data. We were told that if we didn't act quickly, it would be too late for us all, and our window of opportunity to avoid millions of deaths would close.

In the case of the pandemic, this tactic was used to justify the widespread use of novel therapies never before used on a global scale.

In the case of climate change, we're being told we have a matter of only a few years to form a new economy, reduce our travel, change what we eat and how our food is grown, and convert to an electrical-based system of energy and transportation relying upon the use of 17 highly toxic rare-earth elements that are located in a handful of finite deposits on our planet . . . or the whole world will end.

In the United States, we have literally heard from our leaders warning us that we have fewer than 10 years of life remaining if we do not adopt extreme policies being presented to us, and given the urgency of the situation, that there is no time to discuss the facts or develop additional options.

This is a textbook example of the "ensure there's no way out" tactic for indoctrination of a population.

## The Remedy for "Ensure There's No Way Out"

Whether we're making choices regarding a climate crisis, a viral pandemic, or another kind of emergency, our decisions must be based upon information, and the information we want comes from what the scientific community tells us. Unless we are trained as geologists, geophysicists, climate scientists, or infectious disease specialists, most of us are not sufficiently well versed in the specifics of these disciplines to assess research studies on our own. And this is the point. We need to receive input from many scientists with varied perspectives on issues when it comes to forming opinions and making decisions that will affect our bodies and our planet's future.

The reason one scientist is not enough is that science is not a static discipline. Science is a living and dynamic body of information. It is made to be revised and updated as new information comes to light. And it's precisely for this reason that we need to be especially careful anytime the leader of an organization, or the leader of a nation, declares that "the discussion is over" or the science is "settled." This kind of language is designed to shut down any conversation that might lead to viable and healthy solutions that do not fit the vision or the agenda that we are being pressured to support.

The remedy for indoctrination tactic 3 is to ask yourself, "Is this true?" "Is it true that the world will end in 10 years if we don't adopt a radical agenda that upsets the economic balance of our civilization and destroys the livelihoods and lifestyles of millions of people?" "Is it true that the 200,000 years of evolutionary fine-tuning of our immune system will fail us when it encounters a novel virus engineered in a laboratory?" The only way to answer these questions is through knowledge.

We owe it to ourselves to, at the very least, have an elementary understanding of our body and our immune system, and to know what it takes to care for both. We owe it to ourselves to have a basic understanding of Earth's history, geological cycles, and the climate in the past. When we meet someone in a position of power and authority who tells us that there is only one solution to any problem, we owe it to ourselves—and them—to ask, "Is this true?"

More often than not, we will discover that the ability to ask and answer this question objectively will reveal a universal truth: there is always a choice, and always a solution. It may not be the choice or the solution that we imagined utilizing in the past. The solution available to us may not lead to the outcome that we've expected. But there is always an option. And it's this fact that helps us to recognize when we're faced with this tactic of indoctrination.

Recognizing the tactics of indoctrination is a powerful tool that allows us to discern the choices available to us when we're being coerced into accepting agendas that may not be in our best interests. Once we do so, it is our powers of divinity—our ability to transcend perceived limitations—that lead to the insights that provide options and solutions to the challenges that we face.

**PURE HUMAN TRUTH 66:** Tactics used to coerce us into accepting AI and digital technology in our lives and bodies include false narratives about our reality. Fear, guilt, and the belief that we have no other options will be levers used to pressure us to conform.

## THE ACTION

Before his death in 1983, theorist and futurist Buckminster Fuller may have given us the best solution to addressing the principles of indoctrination and ending the Dangerous Game. He stated clearly and wisely: "You never change things by fighting the existing reality. To change something, build a new model that makes the existing model obsolete."[28]

So, what do we do to transcend the threat of transhumanism? The answer is as simple as the game itself. The Dangerous Game that's destroying our humanness, families, and nations can only exist as long as there are players—as long as we agree to support the game.

- When we stop relying on other people to tell us what and how to think, the game cannot exist.
- When we think critically about issues rather than automatically going along with the group thinking in our community, the game cannot exist.
- When we say no to the addictive lure of scanning social media feeds for the news of the day that merely reflects our preexisting opinions, the game cannot exist.

- When we search for the facts rather than relying upon a skewed, abbreviated or AI-altered video of something that triggers us, the game cannot exist.

**PURE HUMAN TRUTH 67:** Our indoctrination can only be effective, and the Dangerous Game can only continue, if we choose to accept indoctrination tactics in our lives.

What would happen if the instigators of the Dangerous Game called for a riot in a big city or for an army to start a war in a country half a world away and no one showed up? In a similar sense, what would happen if we (as well as our families, friends, and neighbors) chose to take responsibility for our personal health, and designed lifestyles that effectively supported us in having strong, natural immunity without needing to inject synthetic chemical substances into our bodies?

And what could happen if we awaken the soft technology of the neurons in our hearts as well as our brain and began to live according to the deeper truth of what it means to be human?

The answer to these questions is simple. When we choose these things, the programming ends. The indoctrination fails, and the Dangerous Game is over.

It's then that we heal and begin anew.

CHAPTER EIGHT

# Pure Human

## The Keys

Who are we . . . but the stories we tell about ourselves, particularly if we accept them?

— Scott Turow (1949– ), American author and lawyer

In December 1945, a discovery was made that sent shock waves through the world of biblical and religious scholarship. It was near the town of Nag Hammadi on the West Bank of Egypt's upper Nile River that a young boy unearthed a mysterious earthen vessel. It had been sealed for centuries and contained the remains of 13 bound codices (ancient texts that were bound as books) as well as loose papyrus pages. Now known as the Nag Hammadi Library, these documents were believed to date to a time between the 2nd and 4th centuries C.E., and they were recognized as early Gnostic texts.

The shock waves in the academic community resulted from the fact that the books were records of biblical writings that had never before been seen, accounts that were very different from the same texts found in the New Testament of today's Christian Bible. In addition to familiar gospels, scholars discovered the remains of additional texts that the early Christian Church chose to exclude when the Bible was being compiled by the various councils and popes of the early Catholic Church in the 4th century.

Among the Gnostic texts revealed were the numerous missing or "lost" books of the Bible, with titles that included the Gospel of Truth, the Origin of the World, the Secret Book of James, the Secret

Book of John, and the Sophia of Jesus Christ, as well as The Thunder, Perfect Mind. This last lost book is now believed to be one of the few remaining texts written by a Gnostic woman. Perhaps the most widely known and controversial of the Nag Hammadi texts is the mysterious Gospel of Thomas.

This gospel is attributed to Thomas Didymus, who is believed to have been the biological brother of Jesus of Nazareth and served as his scribe. Thomas recorded 114 sayings that are attributed directly to Jesus from his sermons. Included in the wisdom of this text is saying number 70, which reads: "If you bring forth what is within you, what you bring forth will save you. If you do not bring forth what is within you, what you do not bring forth will destroy you."[1]

The question that has plagued scholars since the discovery of this gospel is this: What is it that we have within us that has the power to both save us if it's acknowledged, or to destroy us if it is ignored?

In a way that is similar to the message that George Gurdjieff received from the abbot of the secret monastery of the Sarmoung Brotherhood, when he was encouraged to "stay here until you acquire a force in you that nothing can destroy," the messages Thomas recorded reveal a deep and sacred truth that has driven humans to search the world over in hopes of finding it. Both teachings prompt us to evoke something that we already have within us—a force that nothing can destroy, yet one that will potentially destroy us if we deny its existence. I can't help but believe that it's our divinity that is the force described in both teachings.

Today, we find ourselves having come full circle from a kind of thinking that was popular in the era when the Gnostic gospels were discovered in the mid-20th century. In the 1940s, modern science was only beginning to suspect the power we have at our fingertips resulting from weapons technology that was developed during World War II and our ability to engineer the human body.

Today, less than a century later, the seeds of technological power that were initially planted by the 20th-century discoveries have matured to become the advanced weapons and transhumanistic technology of the 21st century. In an ironic twist of fate, however, the same inventions that were initially developed to protect us from our enemies and save our species from suffering and disease have now become the technologies that now threaten our world, our species, and our very existence.

It's up to us to chart the course corrections that have placed us on the destructive trajectory where we find ourselves today. It's up to us to write the pages of the new human story as we claim our destiny as pure humans.

## OUR DESTINY

Two hundred thousand years ago, something miraculous happened here on Earth, something so extraordinary that it left us with god-like abilities that had never before and have never since been given to any other form of life—at least not to any life-form that we're currently aware of.

It was at that time that we were imbued with the power to self-regulate our consciousness and self-heal physically as well as the ability to do so consciously and on demand. We're the only form of life on Earth that can access higher brain states and advanced realms of deep intuition intentionally, when we choose to do so.

We're the only form of life that can create a super immune response and super resilience to our rapidly changing world, awaken longevity enzymes, and create elevated states of super learning, super memorization, super cognition, and more—all at the time and place of our personal choosing.

The DNA evidence shows us that our anatomically modern human ancestors were given these extraordinary abilities, and many more, on the day they mysteriously appeared on the earth. These abilities have been with us every day since that time, and they remain with us today. Although we're seldom informed or reminded of our exceptional abilities that give us deeper meaning to our lives, they are part of us, nonetheless. It's the undeniable fact of these extraordinary powers that points us to our inevitable destiny: we're a species meant to express our powers and live the truth of our human divinity. We're wired to awaken the innate abilities that we've thought of in the past as supernatural powers. And we're wired to embrace these abilities as a natural part of our everyday lives.

**PURE HUMAN TRUTH 68:** Ten thousand generations ago, humans appeared on Earth with extraordinary and godlike abilities given to no other form of life that we know of today.

When we embrace our specialness—the power of our divinity—we give ourselves the reasons to think differently about ourselves, one another, and our relationship to the world around us. We feel empowered to trust our abilities of healing and intuition as they inform us through our choices and our decision-making processes. When we accept our divinity, we learn to trust, and to rely upon, our intuition for everything from managing the details of business and our livelihood to resolving the challenges that inevitably arise in our most intimate relationships.

To accept our divinity and embrace the humanness that makes our divinity possible in this world means we are leveling the playing field in every aspect of our lives. We do so by awakening our ability to transcend our perceived limitations, which is the very definition of our divinity. In living the fullness of our innate capacities, we ensure the future of our species and mark the course correction for the new human story.

As we stand in the fullness of our power, it is impossible for us to be deceived by the dark forces that are sweeping through our societies and across the nations of the Earth today. In the fullness of our power, we can no longer be reduced to the status of powerless victims in the battle between good and evil.

As we learn to trust our divinity, we will also awaken our natural desire for the freedom that leads us to become the best versions of ourselves and to create the best world possible. The key to living our pure human destiny is that we must choose to embrace it.

## THE CHOICE

Unless we choose to think differently about ourselves and accept the truth of our divinity into our lives, the best minds of our time tell us that we're clearly on a course that will bring our worst fears of a world inhabited by an emotionless and AI-driven hybrid species to fruition. Unless we openly claim our humanness and the values that we cherish in our species as sacred, and refuse to allow others to diminish them, we'll find ourselves in a world reflecting the dark agendas of a handful of individuals whom most people have never heard of and don't even know exist.

The question that we must ask of ourselves is simple. It's all about love. Do we love ourselves enough to preserve the gift of our humanness, and choose our human destiny, over the illusion of progress and efficiency that is presented to us through advanced technology? Ultimately, the deeper question becomes, do we love ourselves enough to accept the divinity that is made possible through our humanness, and the power of fearless love and deep healing that it holds for our lives? How we answer this question will either lead us to the heights of our greatest godlike destiny or to the depths of our darkest fear-based fate.

**PURE HUMAN TRUTH 69:** Will we choose the gift of our humanness or succumb to the efficiency of becoming technologically enhanced forms of life?

If we accept the evidence suggesting that advanced beings have traveled from distant worlds to warn us of the consequences of replacing our humanness with technology, then maybe their visits give us a clue as to where we go from here. Both their message and the discoveries of the best science of our time lead to the same conclusion.

The message and the science remind us that we're worth preserving. Even with the dysfunctional expressions of war and hate that we face today, something remains in our humanness that is so rare, so precious, and so beautiful that it would be an evolutionary tragedy to let that something, and us, slip away.

This book is written with one purpose in mind: to give a voice to the story of our potential and to empower us in the acceptance of our specialness. The power of the new human story holds the key to ending the fear and the hate that divides us and to making the choices that lead to healthy and thriving lives in a world transformed.

**PURE HUMAN TRUTH 70:** The choice is ours to make.

## THE PURE HUMAN TRUTHS

The following Pure Human Truths, which encapsulate the information in this book, give us reasons to cherish our humanness and to remember what a rare, precious, and beautiful form of life we are.

**PURE HUMAN TRUTH 1:** For the first time in human history, we're implementing technology that irreversibly changes our bodies on a biological level.

**PURE HUMAN TRUTH 2:** By the year 2030, we will either have awakened to the truth of our untapped human potential, or we will be locked into a society of hybrid humans that has engineered away our powers of creativity, emotion, empathy, and intuition.

**PURE HUMAN TRUTH 3:** We stand at the precipice of giving away our humanness—the biological bridge to our divinity.

**PURE HUMAN TRUTH 4:** In many respects, human cells and specialized neurons are superior in performance, scalability, and adaptability to hardwired computer chips and the limited algorithms of AI.

**PURE HUMAN TRUTH 5:** We owe it to ourselves to recognize the deep truth of what it means to be human before we give ourselves away to the technology now being proposed by the transhumanist movement.

**PURE HUMAN TRUTH 6:** Divinity is defined as powers, or forces, that transcend perceived limitations.

**PURE HUMAN TRUTH 7:** The battle between good and evil is ultimately the battle for human divinity.

**PURE HUMAN TRUTH 8:** Divinity is the part of us that is ancient and timeless, where our direct knowing, imagination, creativity, self-acceptance, and self-healing begin.

**PURE HUMAN TRUTH 9:** Expressing our divinity frees us from the fear that keeps us feeling small, insignificant, and powerless, allowing us to triumph over life's challenges.

**PURE HUMAN TRUTH 10:** Awakening your divinity begins with the way you think of yourself—your story.

**PURE HUMAN TRUTH 11:** We're more than the result of random processes. It is statistically beyond chance that the seven Goldilocks conditions that make our world and our lives possible are the result of "lucky" physics.

**PURE HUMAN TRUTH 12:** Our universe appears to be alive, conscious, and intelligent.

**PURE HUMAN TRUTH 13:** The first of our kind appeared on Earth approximately 200,000 years ago, we're still here, and our DNA blueprint hasn't changed.

**PURE HUMAN TRUTH 14:** It is mathematically impossible that the mutations responsible for our most cherished human qualities, including empathy and intuition, are the result of random processes and "lucky" biology.

**PURE HUMAN TRUTH 15:** In 2007, the first message was written into, and later retrieved from, a living organism, proving that it's possible to write, store, and retrieve intelligent information in DNA.

**PURE HUMAN TRUTH 16:** DNA is more efficient as a storage medium than computer flash storage, making it a likely candidate for the location of an ancient message coded into our biology long ago.

**PURE HUMAN TRUTH 17:** Human cells may be thought of from an information technology perspective, with each cell being a library, the chromosomes being books, and the shorter strands of genes being chapters, paragraphs, sentences, and words.

**PURE HUMAN TRUTH 18:** The characters of cuneiform, Sanskrit, Arabic, and Hebrew have numeric equivalents that may be used interchangeably in written texts.

**PURE HUMAN TRUTH 19:** The unique numbers that represent the atomic mass of each of the four elements that compose our DNA bases can be added up to total numbers that form various words written in the ancient root languages.

**PURE HUMAN TRUTH 20:** When we perform the substitution of letters for their numeric equivalents from the atomic mass of our DNA, the first layer of the code for each cell of our bodies reads as: "God/ eternal within the body."

**PURE HUMAN TRUTH 21:** The statistical odds of the message "God/ eternal within the body" forming within our DNA by chance are .00042 percent, telling us that the probability of us carrying a message of such significance is beyond chance.

**PURE HUMAN TRUTH 22:** When we become discouraged, distracted, or deceived, or we simply have forgotten who we are, we need look no further than the 50 trillion cells within our bodies to be reminded that we are literally God/eternal within the body.

**PURE HUMAN TRUTH 23:** Transhumanism is a philosophy that advocates incorporating AI, computer chips, and electronic sensors into the human body to "fix" the flaws of our natural biological functions.

**PURE HUMAN TRUTH 24:** Although the first successful gene editing of two human embryos in 2018 was illegal, it demonstrated that the process is beyond theory. It is now possible to successfully engineer human DNA in the womb after conception.

**PURE HUMAN TRUTH 25:** There are three stages of transhumanism ranging from prosthetic replacement for the body to the proposed capture and storage of an individual's consciousness on a computer chip.

**PURE HUMAN TRUTH 26:** The inability to clone any form of life that survives its natural lifespan demonstrates that there is something missing in the cloning model, something that is not accounted for in the current thinking when it comes to consciousness and life.

**PURE HUMAN TRUTH 27:** The DNA in the nucleus of a cell must communicate with the DNA that is outside of the nucleus to successfully find resonance with the information that leads to a successful life.

**PURE HUMAN TRUTH 28:** One of the dangers of transhumanism is that when we replace our natural biology with artificial technology, the natural functions begin to weaken and atrophy.

**PURE HUMAN TRUTH 29:** Some Indigenous traditions suggest that we build the complex world of machines and technology outside of us to remind ourselves that they mimic the abilities that already live inside of us.

**PURE HUMAN TRUTH 30:** Through our art, technology, books, music, and films, we communicate with ourselves about the things that we are asking ourselves to remember.

**PURE HUMAN TRUTH 31:** The most popular and successful movies of our day are those that portray humans remembering or discovering hidden talents and superpowers.

**PURE HUMAN TRUTH 32:** We are a sophisticated soft technology with the functions of our cells meeting and, in some cases, exceeding the capability of the functions of AI and the components on a computer chip.

**PURE HUMAN TRUTH 33:** In addition to our cells functioning like electrical components, our DNA preserves a record of the genetic transactions that make our species as it is, a blockchain-like record that is transparent, permanent, and immutable.

**PURE HUMAN TRUTH 34:** The capacity and function of a human neuron is adaptable and scalable for performance beyond any known limits rather than being limited in the way that physics limits the capacity of solid-state computer chips.

**PURE HUMAN TRUTH 35:** We self-regulate our advanced soft technology through the user interfaces of thought, feeling, emotion, breath, and other epigenetic factors.

**PURE HUMAN TRUTH 36:** In the year 2000, scientists discovered that the human genome is made of only about 24,000 genes, meaning that the previously accepted relationship of one gene producing each of the roughly 100,000 proteins in the human body was wrong.

**PURE HUMAN TRUTH 37:** Through epigenetics, it is possible for one human gene to "program" up to 100 unique proteins. This means we can do more programming with a few highly efficient genes.

**PURE HUMAN TRUTH 38:** Mirror neurons do not know the difference between having a lived experience and watching someone else have an experience.

**PURE HUMAN TRUTH 39:** The way we see ourselves in our imagination triggers our mirror neurons to signal the body with the information to support our self-image.

**PURE HUMAN TRUTH 40:** A neural network, or "little brain," within the human heart thinks, feels, experiences, and remembers independently of the cranial brain.

**PURE HUMAN TRUTH 41:** We are the only known form of life that can intentionally harmonize the neural networks of its heart and brain to create an optimized state of heart-brain coherence.

**PURE HUMAN TRUTH 42:** The pure human body is a highly advanced, technologically sophisticated soft technology that, in many ways, outperforms artificial components of computer technology to make possible advanced states of consciousness and healing.

**PURE HUMAN TRUTH 43:** The greater our expression of divinity, the more of our destiny we achieve.

**PURE HUMAN TRUTH 44:** We are in constant, resonant communication with the world around us through the biological antennae of our molecules and cells.

**PURE HUMAN TRUTH 45:** When we witness one person succeed where others have failed, their accomplishment is a bridge that makes it easier for us to follow in their footsteps.

**PURE HUMAN TRUTH 46:** When we witness another person transcend limitations that we've accepted for ourselves, we must choose either to discount what we've seen, viewing it as unattainable, or to accept what we've seen and shift our belief system to accommodate the new belief.

**PURE HUMAN TRUTH 47:** To live our divinity is to heal the deceptive belief that we are flawed, frail, and powerless beings who need something outside of ourselves to succeed in the world and thrive in life.

**PURE HUMAN TRUTH 48:** In 2020, the World Economic Forum announced plans for the Great Reset, an unprecedented attempt to build a new digital society in the wake of the global pandemic that was still underway then and the chaos of the lockdowns occurring in many countries that year.

**PURE HUMAN TRUTH 49:** The vision for the Great Reset is to merge us and all the present-day systems of finance, business, manufacturing, transportation, and food production that we use, as well as our consumption, travel, lifestyle choices, and spending habits, into one vast network that is managed and regulated through the oversight of an advanced AI.

**PURE HUMAN TRUTH 50:** The key to the success of the Great Reset is the digital merging of humans, including our intimate biometrics and vital signs, as well as signs of anger, fear, and joy, into the global matrix to be interpreted by automated systems.

**PURE HUMAN TRUTH 51:** In 2019, the United Nations formally entered a partnership with the World Economic Forum to expedite the implementation of the Great Reset using the 2030 Sustainable Development Goals as the vehicle to do so.

**PURE HUMAN TRUTH 52:** Speaking at Davos to the organizations, corporations, and financial institutions whose operations and purposes impact our daily lives, Yuval Noah Harari declared that human biology is now a hackable "technology."

**PURE HUMAN TRUTH 53:** Transhumanists view some of our most cherished characteristics and abilities as flaws that can be remedied and "fixed" using the advanced technology of gene editing and nanorobotics available today.

**PURE HUMAN TRUTH 54:** Using gene editing and chemical therapies to manage human conception and births to ensure a predetermined outcome of "desirable" traits is the high-tech version of the old philosophy of eugenics.

**PURE HUMAN TRUTH 55:** The philosophy of eugenics and selective breeding is inherent in the transhumanistic vision of the Great Reset.

**PURE HUMAN TRUTH 56:** Younger generations are especially vulnerable to the threats of transhumanism because they've grown up in a technological world where the lure of computers and AI is touted as humanity's saving grace.

**PURE HUMAN TRUTH 57:** A common thread that is documented among the reports of alien abductions throughout the world is that the abductors are warning us of the consequences of choosing a transhumanistic path of human evolution.

**PURE HUMAN TRUTH 58:** The inherent danger of choosing the transhumanistic path is that the changes engineered into the human genome, once made, cannot be reversed.

**PURE HUMAN TRUTH 59:** The new discoveries of human biology and its divinity now give us everything we need to preserve our evolutionary heritage and make healthy choices that honor our humanness.

**PURE HUMAN TRUTH 60:** The vision of the Great Reset requires us to accept technology into our lives and into our bodies in order that we may become integrated into a large-scale digital landscape.

**PURE HUMAN TRUTH 61:** The power, imagination, and freedom of human divinity are seen as obstacles to those who are attempting to achieve the goals of the Great Reset.

**PURE HUMAN TRUTH 62:** Under the pretext of providing us an easier life, we are being led to ask for the very changes in society and our lives that are destroying the fabric of society and our lives.

**PURE HUMAN TRUTH 63:** Fifth-generation warfare plays out in the battlefield of our minds before there is ever a kinetic engagement between people.

**PURE HUMAN TRUTH 64:** The goal of the Dangerous Game is to be led to diminish one another—to destroy the trust and social bonds that hold together our most intimate relationships, our families, and our societies until we are led to question and doubt our own value and humanness.

**PURE HUMAN TRUTH 65:** Through the Dangerous Game, differences in race, religion, culture, and diversity that we have celebrated as strengths in the past are weaponized, leaving us divided and vulnerable to the ideas and agendas of others.

**PURE HUMAN TRUTH 66:** Tactics used to coerce us into accepting AI and digital technology in our lives and bodies include false narratives about our reality. Fear, guilt, and the belief that we have no other options will be levers used to pressure us to conform.

**PURE HUMAN TRUTH 67:** Our indoctrination can only be effective, and the Dangerous Game can only continue, if we choose to accept indoctrination tactics in our lives.

**PURE HUMAN TRUTH 68:** Ten thousand generations ago, humans appeared on Earth with extraordinary and godlike abilities given to no other form of life that we know of today.

**PURE HUMAN TRUTH 69:** Will we choose the gift of our humanness or succumb to the efficiency of becoming technologically enhanced forms of life?

**PURE HUMAN TRUTH 70:** The choice is ours to make.

# Acknowledgments

*Pure Human* marks my 10th book as a Hay House author. My writing of the book, however, was only the beginning of the process. Through a cooperative effort that most readers will never see, a dedicated community of copy editors, proofreaders, and graphic designers; social media, marketing, and publicity strategists; event producers, sales representatives, book distributors, and bookstore managers had to arrange their lives and their schedules around my promise that *Pure Human* would be written and delivered when I promised.

While I will never meet most of this community personally, I know they're there, and I'm deeply grateful and eternally thankful for all that they do each day to share the information, insights, techniques, and human stories that make this world a better place. I would like to take this opportunity to express my gratitude to those whose efforts have directly contributed to making this book possible. Specifically, I want to express my gratitude to:

Louise Hay, for her unwavering belief in our potential to heal and to love ourselves into healing, and for expressing her vision for the extraordinary family that has become the remarkable publishing company Hay House, Inc. Though Louise left this world in 2017, her intuitive philosophy laid the foundation for writing and publishing *Pure Human*.

Reid Tracy, president and CEO, Hay House, for your vision and personal dedication to the truly extraordinary way of doing business that has become the hallmark of Hay House's success, and especially for your support and rock-solid advice, and your trust in me and my work since our first meeting in 2003.

Margarete Nielsen, COO, Hay House, for your vision, dedication, and leadership. I'm especially grateful for your sage advice that opens a window from my personal desk in New Mexico into the big, ever-changing world of media and publishing, your trust in me and my decisions, and your always-present friendship and support.

Patty Gift, vice president, publisher, Hay House. Who could have known when you introduced me to Harmony Books in 1999 where our journey would lead? Thank you for your trust, advice, wisdom, and support over two decades of life shifts and world changes. Most of all, thank you for your unwavering friendship.

Anne Barthel, executive editor, Hay House. I'm honored and blessed to know you as my most amazing and talented all-around literary guru, most awesome editor, and trusted sounding board, and now as my dear friend.

In addition, I am grateful to Ned Leavitt, my one and only literary agent. Thank you so very much for your wisdom, your integrity, and the human touch that you bring to every milestone we cross together. Through your guidance in shepherding our books through the ever-changing world of publishing, we have reached countless numbers of people in over 70 countries on six continents with our empowering message of hope and possibility. While I deeply appreciate your impeccable guidance, I am especially grateful for your trust in me and our friendship.

My heartfelt gratitude and deepest appreciation for Stephanie Gunning, my first-line editor extraordinaire since the beginning. Together, we've covered a lot of ground to share the eternal message of our human potential, and through it all, you're still my dear friend. You have my deepest respect for your knowledge of the world, your impeccable language skills, your ability to treat each of our books as if it's our first, and my gratitude for the way you generously shower your expertise and experience onto each of our projects.

I am proud to be part of the virtual team, and the family, that has grown around the support of my work over the years, including Lauri Willmot and Chelsey Luikart. Lauri, my dear friend and confidant since 1996 and now the executive director of our company, Wisdom Traditions. I admire your strength, wisdom, and clear thinking, and I respect you deeply and appreciate the countless ways that you're there for me always, especially when it counts. I look forward to the new journey that we've embarked upon and the mystery of where it will lead us. You can't retire until I do!

Chelsey, your skills and expertise have opened a new dimension to Wisdom Traditions and how we can effectively share what we do with new, younger, and ever-growing audiences. I'm deeply grateful for all of the hats you wear to keep me and our company on track and on time for the many commitments we promise each month, and for keeping our events, marketing, and media current. We'll be adding the new addition to your family to our team before you know it!

Thank you, Rita Curtis, my bookings manager extraordinaire, and now my friend: I deeply appreciate your vision, your clarity, and your skills that get us from here to there each month. Most of all, I appreciate your trust, your openness to new ideas, and especially our growing friendship.

To my beautiful wife, Martha, thank you for your lasting friendship, gentle wisdom, and all-embracing love that is with me each day of my life. Along with the furry beings with whom we share our lives, you are the family that believes in me each day and gives me the reason to come home from each event. Thank you for all that you share and bring to my life.

A very special thanks to everyone who has supported my work, books, recordings, and live presentations over the years. I am honored by your trust, in awe of your vision for a better world, and deeply appreciative of your passion for bringing that world into existence. Through your presence, I have learned to become a better listener, and heard the words that allow me to share our empowering message of hope and possibility. To all, I remain grateful in all ways, always.

# Endnotes

## INTRODUCTION


1. Ray Kurzweil, "Ray Kurzweil—Immortality by 2045," 2045 Initiative, March 4, 2013, YouTube video, https://www.youtube.com/watch?v=f28LPwR8BdY&t=38s.

2. Ray Kurzweil, *The Singularity Is Near: When Humans Transcend Biology* (New York: Viking, 2005): 136.


## CHAPTER ONE: We Are the Prize


1. "Divinity," Wikipedia, accessed July 11, 2024.

2. Ned Herrmann, "What Is the Function of the Various Brainwaves?" *Scientific American*, December 22, 1997, https://www.scientificamerican.com/article/what-is-the-function-of-t-1997-12-22.

3. Rollin McCraty, Mike Atkinson, and Raymond Trevor Bradley, "Electrophysiological Evidence of Intuition: Part 1. The Surprising Role of the Heart," *Journal of Alternative and Complementary Medicine* 10, no. 1 (February 2004): 33–43, https://doi.org/10.1089/107555304322849057.

4. Asad Meah, Quote 4 in "15 Inspirational Quotes on the Superconscious Mind," October 17, 2023, Awaken the Greatness Within, https://www.awakenthegreatnesswithin.com/15-inspirational-quotes-on-the-superconscious-mind.

5. "Hazrat Inayat Khan Quotes," QuoteFancy.com, accessed July 11, 2024, https://quotefancy.com/hazrat-inayat-khan-quotes.

6. "Hazrat Inayat Khan Quotes."

7. Julie Ribaudo et al., "Maternal History of Adverse Experiences and Posttraumatic Stress Disorder Symptoms Impact Toddlers' Early Socioemotional Wellbeing: The Benefits of Infant Mental Health-Home Visiting," *Frontiers in Psychology* 12 (January 17, 2022): 792989, https://doi.org/10.3389/fpsyg.2021.792989.

8. Alison Schafer, "Syria's Children—How Conflict Can Harm Brain Development," World Vision, January 6, 2014, https://www.wvi.org/experts/article/syria%E2%80%99s-children-%E2%80%93-how-conflict-can-harm-brain-development.

9. Vicky Stein, "Goldilocks Zone: Everything You Need to Know about the Habitable Sweet Spot," Space.com, February 16, 2023, https://www.space.com/goldilocks-zone-habitable-area-life.

10. Robert H. Dicke, "Dirac's Cosmology and Mach's Principle," *Nature* 192 (November 4, 1961): 440–1, https://doi.org/10.1038/192440a0.

11. Duane Elgin, "Why We Need to Believe in a Living Universe," *Huffington Post*, May 15, 2011, updated July 16, 2011, http://www.huffingtonpost.com/duaneelgin/living-universe_b_862220.html.

12. Gregory L. Matloff, "Panpsychism as an Observational Science," *Journal of Consciousness Exploration and Research* 11, no. 5 (August 2020): 468–86, https://jcer.com/index.php/jcj/article/view/900/911.

13. Gregory L. Matloff, "Can Panpsychism Become an Observational Science?" *Journal of Consciousness Exploration and Research* 7, no. 7 (August 2016): 524–43, https://jcer.com/index.php/jcj/article/view/579/595.

14. "Acceptance Address by Prof. Freeman Dyson," TempletonPrize.org, May 16, 2000, https://www.templetonprize.org/laureate-sub/dyson-acceptance-address.

15. Francis Reddy, "NASA Missions Study What May Be a 1-in-10,000-Year Gamma-Ray Burst," NASA, March 28, 2023, https://www.nasa.gov/universe/nasa-missions-study-what-may-be-a-1-in-10000-year-gamma-ray-burst.

16. Gregory L. Matloff, "Stellar Consciousness: Can Panpsychism Emerge as an Observational Science?" *EdgeScience* 29 (March 2017): 9–14, https://docslib.org/doc/2819479/stellar-consciousness-can-panpsychism-emerge-as-an-observational-science.

17. Elgin, "Why We Need to Believe."

18. Ray Bradbury, "G. B. S.—Mark V," *I Sing the Body Electric! And Other Stories* (New York: Perennial Books, 1998), 280.

19. Jacob W. Ijdo et al., "Origin of Human Chromosome 2: An Ancestral Telomere-Telomere Fusion," *Proceedings of the National Academy of Sciences of the United States of America* 88, no. 20 (October 15, 1991): 9051–5, https://www.ncbi.nlm.nih.gov/pmc/articles/PMC52649.

20. Ijdo et al., "Origin of Human Chromosome 2."

21. "Chromosome 2," Wikipedia, accessed July 11, 2024, https://en.wikipedia.org/wiki/Chromosome_2.

22. Ijdo et al., "Origin of Human Chromosome 2."

## CHAPTER TWO: Who Are We?


1. S. Mohan, S. Vinodh, and F. R. Jeevan, "Preventing Data Loss by Storing Information in Bacterial DNA," *International Journal of Computer Applications* 69, no. 19 (May 2013): 53–7, https://research.ijcaonline.org/volume69/number19/pxc3888322.pdf.

2. George M. Church, Yuan Gao, and Sriram Kosuri, "Next-Generation Digital Information Storage in DNA," *Science* 337, no. 6102 (August 16, 2012): 1628, https://doi.org/10.1126/science.1226355.

3. Darshan Panda et al., "DNA as a Digital Information Storage Device: Hope or Hype?" *3 Biotech* 8, no. 5 (May 4, 2018): 239, https://doi.org/10.1007/s13205-018-1246-7.

4. "The Thirty Two Rules of Eliezer," Nazarene Judaism, accessed July 11, 2024, https://nazarenejudaism.com/?page_id=105.

5. American Heritage Dictionary of the English Language, 5th ed. (2022), s.v. "science."

6. Benjamin Blech, *The Secrets of Hebrew Words* (New York: Rowman & Littlefield, 1977), 129–32.

7. Willis Barnstone, ed., *The Other Bible* (San Francisco: HarperSanFrancisco, 1984), 25.

8. Peter Taylor, "The Art of Gematria," Provincial Grand Lodge of Forfarshire, 2020, https://www.pglforfarshire.org/The_Art_of_Gematria_PT.html.

9. King James Version.

10. King James Version.


## CHAPTER THREE: Transhumanism


1. Mark O'Connell, quoted in Robin McKie, "No Death and an Enhanced Life: Is the Future Transhuman?" *Observer*, May 6, 2018, www.theguardian.com/technology/2018/may/06/no-death-and-an-enhanced-life-is-the-future-transhuman.

2. J. B. S. Haldane, *Daedalus, or Science and the Future* (London: Kegan Paul, Trench, Trubner, 1923), 44, https://jbshaldane.org/books/1923-Daedalus/haldane-1923-daedalus-ocr.pdf.

3. David Cyranoski and Heidi Ledford, "Genome-Edited Baby Claim Provokes International Outcry," *Nature* 563, no. 7733 (November 26, 2018): 607–8, https://doi.org/10.1038/d41586-018-07545-0.

4. Marilynn Marchione, "Chinese Researcher Claims First Gene-Edited Babies," *Associated Press*, November 26, 2018, https://apnews.com/article/ap-top-news-international-news-ca-state-wire-genetic-frontiers-health-4997bb7aa36c45449b488e19ac83e86d.

5. Marchione, "Chinese Researcher Claims First Gene-Edited Babies."

6. Howard Gest, "The July 1945 Szilard Petition on the Atomic Bomb: Memoir by a Signer in Oak Ridge," Department of Biology, Indiana University, accessed July 11, 2024, https://biology.indiana.edu/documents/historical-materials/gest_pdfs/hgSzilard.pdf.

7. Abhijit Naskar, *Mission Reality* (independently published, 2019): 75.

8. Markus Aldén et al., "Intracellular Reverse Transcription of Pfizer BioNTech COVID-19 mRNA Vaccine BNT162b2 in Vitro in Human Liver Cell Line," *Current Issues in Molecular Biology* 44, no. 3 (February 25, 2022): 1115–26, https://doi.org/10.3390/cimb44030073.

9. Description of Neuralink, its goals, and outreach for volunteers to test the technology. Neuralink.com, accessed July 11, 2024, https://neuralink.com.

10. Tim Levin, "Elon Musk Wants to Give Amputees Robotic Limbs Powered by Chips Implanted in Their Brains," *Business Insider*, July 20, 2023, https://www.businessinsider.com/elon-musk-optimus-tesla-robot-limbs-neuralink-cyborg-2023-7.

11. Jordan Inafuku et al., *Downloading Consciousness* (online information series), Stanford University Department of Computer Science, accessed July 11, 2024, https://cs.stanford.edu/people/eroberts/cs201/projects/2010-11/DownloadingConsciousness/tandr.html.

12. Susan Schneider, "Merging with AI Would Be Suicide for the Human Mind," *Financial Times*, August 13, 2019, https://www.ft.com/content/0c4fac58-bd15-11e9-9381-78bab8a70848.

13. Greg Egan, *Axiomatic: Short Stories of Science Fiction* (New York: Night Shade Books, 2014), 167.

14. Schneider, "Merging with AI Would Be Suicide."

15. Schneider, "Merging with AI Would Be Suicide."

16. Judy Jones. "Cloning May Cause Health Defects," *BMJ* 318, no. 7193 (May 8, 1999): 1230, https://www.ncbi.nlm.nih.gov/pmc/articles/PMC1115633.

17. Jones, "Cloning May Cause Health Defects."

18. Jones, "Cloning May Cause Health Defects."

19. Rupert Sheldrake, "Morphic Resonance and Morphic Fields—an Introduction," Sheldrake.org, accessed July 11, 2024, https://www.sheldrake.org/research/morphic-resonance/introduction.

20. Pushpendra Singh et al., "DNA as an Electromagnetic Fractal Cavity Resonator: Its Universal Sensing and Fractal Antenna Behavior," in *Soft Computing: Theories and Applications: Proceedings of SoCTA 2016*, Advances in Intelligent Systems and Computing 584 (Singapore: Springer), 2:213–23, https://doi.org/10.1007/978-981-10-5699-4_21.

21. T. J. Shors et al., "Use It or Lose It: How Neurogenesis Keeps the Brain Fit for Learning," *Behavioural Brain Research* 227, no. 2 (February 14, 2012): 450–8, https://doi.org/10.1016/j.bbr.2011.04.023.

22. Bart Ellenbroek and Jiun Youn, "Rodent Models in Neuroscience Research: Is It a Rat Race?" *Disease Models & Mechanisms* 9, no. 10 (October 1, 2016): 1079–87, https://doi.org/10.1242/dmm.026120.

23. "Evil," Wikipedia, accessed July 11, 2024.

## CHAPTER FOUR: The Secret

1. Kelly White, https://www.goodreads.com/author/quotes/210231.Kelly_White.

2. "Moore's Law 40th Anniversary with Gordon Moore," Computer History Museum, December 17, 2007, YouTube video, https://www.youtube.com/watch?v=MH6jUSjpr-Q.

3. Shekh M. Mahmudul Islam, "Performances of Multi-Frequency Voltage to Current Converters for Bioimpedance Spectroscopy," *Bangladesh Journal of Medical Science* 5, no. 1 (October 2012): 221–4721, https://doi.org/10.3329/bjmp.v5i1.14671.

4. Beverly Rubik et al., "Biofield Science and Healing: History, Terminology, and Concepts," *Global Advances in Health and Medicine* 4, supplement (November 1, 2015): 8–14, https://doi.org/10.7453/gahmj.2015.038.suppl.

5. Pushpendra Singh et al., "DNA as an Electromagnetic Fractal Cavity Resonator: Its Universal Sensing and Fractal Antenna Behavior," in *Soft Computing: Theories and Applications; Proceedings of SoCTA 2016*, Advances in Intelligent Systems and Computing 584 (Singapore: Springer), 2:213–23, https://doi.org/10.1007/978-981-10-5699-4_21.

6. Martin Blank and Reba Goodman, "DNA Is a Fractal Antenna in Electromagnetic Fields," *International Journal of Radiation Biology* 87, no. 4 (February 28, 2011): 409–15, https://doi.org/10.3109/09553002.2011.538130.

7. David Bohm, *Wholeness and the Implicate Order* (New York: Routledge Classics, 2002).

8. Reza Rastmanesh and Matti Pitkänen, "Can the Brain Be Relativistic?" *Frontiers in Neuroscience* 15, (June 17, 2021): 659860, https://doi.org/10.3389/fnins.2021.659860.

9. Naveen Nagarajan and Charles F. Stevens, "How Does the Speed of Thought Compare for Brains and Digital Computers?" *Current Biology* 18, no. 17 (September 9, 2008): R756–8, https://doi.org/10.1016/j.cub.2008.06.043.

10. Nagarajan and Stevens, "How Does the Speed of Thought Compare."

11. Gladys Barragan-Jason et al., "Fast and Famous: Looking for the Fastest Speed at Which a Face Can Be Recognized," *Frontiers in Psychology* 4 (March 3, 2013), https://doi.org/10.3389/fpsyg.2013.00100. Also see: Gabrielle Shea, "Face Recognition Technology Accuracy and Performance," Bipartisan Policy Center, May 24, 2023, https://bipartisanpolicy.org/blog/frt-accuracy-performance.

12. White House Office of the Press Secretary, "June 2000 White House Event," news release, June 26, 2000, updated August 29, 2012, https://www.genome.gov/10001356/june-2000-white-house-event.

13. Giacomo Rizzolatti and Laila Craighero, "The Mirror-Neuron System," *Annual Review of Neuroscience* 27 (July 2004): 169–92, https://doi.org/10.1146/annurev.neuro.27.070203.144230.

14. Flavia Filimon et al., "Human Cortical Representations for Reaching: Mirror Neurons for Execution, Observation, and Imagery," *NeuroImage* 37, no. 4 (October 1, 2007): 1315–28, https://doi.org/10.1016/j.neuroimage.2007.06.008.

15. J. Andrew Armour, *Neurocardiology: Anatomical and Functional Principles* (Boulder Creek, CA: Institute of HeartMath, 2003).

16. Mohamed Omar Salem, "The Heart, Mind and Spirit," Royal College of Psychiatrists, https://www.rcpsych.ac.uk/docs/default-source/members/sigs/spirituality-spsig/spirituality-special-interest-group-publications-professor-mohamed-omar-salem-the-heart-mind-and-spirit.pdf.

17. Rollin McCraty et al., *The Coherent Heart: Heart-Brain Interactions, Psychophysiological Coherence, and the Emergence of System-Wide Order* (Boulder Creek, CA: HeartMath Research Center, Institute of HeartMath, 2009), https://www.heartmath.org/resources/downloads/coherent-heart.

18. "12 Benefits of Brain and Heart Coherence," Alleviant Integrated Mental Health, accessed July 12, 2024, https://alleviant.com/12-benefits-of-brain-heart-coherence.

19. Rollin McCraty et al., "The Impact of a New Emotional Self-Management Program on Stress, Emotions, Heart Rate Variability, DHEA and Cortisol," *Integrative Psychological and Behavioral Science,* 33, no. 2 (April–June 1998): 151–70, https://doi.org/10.1007/BF02688660.

20. Glen Rein, Mike Atkinson, and Rollin McCraty, "The Physiological and Psychological Effects of Compassion and Anger," *Journal of Advancement in Medicine* 8, no. 2 (Summer 1995): 87–105, https://www.heartmath.org/assets/uploads/2015/01/compassion-and-anger.pdf.

## CHAPTER FIVE: Everyday Divinity

1. Meetings with Remarkable Men: Gurdjieff's Search for Hidden Knowledge, directed by Peter Brook (London: Enterprise Pictures, 1979).

2. Kahlil Gibran, *The Prophet* (New York: Alfred A. Knopf, 1963), 28.

3. "Mother's Intuition?" Channel 10 News, May 31, 2012, YouTube video, https://www.youtube.com/watch?v=oMWzsoSBSP8.

4. Amy M. Boddy et al., "Fetal Microchimerism and Maternal Health: A Review and Evolutionary Analysis of Cooperation and Conflict beyond the Womb," *BioEssays* 37, no. 10 (October 2015): 1106–18, https://doi.org/10.1002/bies.201500059.

5. Katya Orlova, "Mother's Day Genetics: How Long Does a Mother 'Carry' a Child?" Ariel Precision Medicine, May 10, 2020, https://arielmedicine.com/mothers-day-genetics-how-long-does-a-mother-carry-a-child.

6. Orlova, "Mother's Day Genetics."

7. Pushpendra Singh et al., "DNA as an Electromagnetic Fractal Cavity Resonator: Its Universal Sensing and Fractal Antenna Behavior," in *Soft Computing: Theories and Applications: Proceedings of SoCTA 2016*, Advances in Intelligent Systems and Computing 584 (Singapore: Springer), 2:213–23, https://doi.org/10.1007/978-981-10-5699-4_21.

8. Luke Chan, "101 Miracles of Natural Healing: Chi-Lel Qigong for Health, Longevity, Creativity, and Mental Clarity" (Beijing: Benefactor, 1995), video.

9. Neville Goddard, *The Power of Awareness* (Marina del Rey, CA: DeVorss, 1952), 10.

10. Russell Targ and Harold Puthoff, "Remote Viewing of Natural Targets," Stanford Research Institute, August 26, 1974, https://www.cia.gov/readingroom/docs/CIA-RDP96-00787R000500410001-3.pdf. Originally classified. Released August 7, 2000.

11. Rollin McCraty, Mike Atkinson, and Raymond Trevor Bradley, "Electrophysiological Evidence of Intuition: Part 1. The Surprising Role of the Heart," *Journal of Alternative and Complementary Medicine* 10, no. 1 (February 2004): pp. 33–43, https://doi.org/10.1089/107555304322849057.

12. Harold E. Puthoff and Russell Targ, "A Perceptual Channel for Information Transfer over Kilometer Distances: Historical Perspective and Recent

Research," *Proceedings of the Institute of Electrical and Electronics Engineers* 64, no. 3 (March 1976): 329–54, https://www.cia.gov/readingroom/docs/CIA-RDP79-00999A000300060005-1.pdf. Originally classified. Released June 24, 2003.

13. Targ and Puthoff, "Remote Viewing of Natural Targets."

14. Mahadeva Srinivasan, "Clairvoyant Remote Viewing: The US Sponsored Psychic Spying," *Strategic Analysis*, 26, no. 1 (January–March 2002), https://ciaotest.cc.columbia.edu/olj/sa/sa_jan02srm01.html.

15. Srinivasan, "Clairvoyant Remote Viewing."

16. Srinivasan, "Clairvoyant Remote Viewing."

17. King James Version.

## CHAPTER SIX: Human or Hybrid?

1. John Mutter, "Opportunity from Crisis: Who Really Benefits from Post-Disaster Rebuilding Efforts," *Foreign Affairs*, April 18, 2016, https://www.foreignaffairs.com/world/opportunity-crisis.

2. Klaus Schwab, "What Is the Fourth Industrial Revolution? by Prof Klaus Schwab," Lee Kuan Yew School of Public Policy, July 13, 2016, YouTube video, https://www.youtube.com/watch?v=7xUk1F7dyvI.

3. "Sustainable Development Goals," World Health Organization, accessed July 12, 2024, https://www.who.int/europe/about-us/our-work/sustainable-development-goals.

4. "Sustainable Development Goals."

5. "Goal 2: Zero Hunger," United Nations, accessed July 12, 2024, https://www.un.org/sustainabledevelopment/hunger.

6. "SDG Target 3.b Essential Medicines and Vaccines," World Health Organization, Global Health Observatory, accessed July 12, 2024, https://www.who.int/data/gho/data/themes/topics/sdg-target-3_b-development-assistance-and-vaccine-coverage.

7. Darshana Narayanan, "The Dangerous Populist Science of Yuval Noah Harari," *Current Affairs*, July 6, 2022, https://www.currentaffairs.org/2022/07/the-dangerous-populist-science-of-yuval-noah-harari.

8. Yuval Noah Harari, "Will the Future Be Human? Yuval Noah Harari," World Economic Forum, January 25, 2018, YouTube video, https://www.youtube.com/watch?v=hL9uk4hKyg4.

9. Yuval Noah Harari, "Will the Future Be Human?"

10. Yuval Noah Harari, "Will the Future Be Human?"

11. "World Economics [*sic*] Forum Discusses mRNA," Tony Heller, January 6, 2023, YouTube video, https://www.youtube.com/watch?v=P8Fn07gMtxY.

12. Stephen Chen, "Chinese Team behind Extreme Animal Gene Experiment Says It May Lead to Super Soldiers Who Survive Nuclear Fallout," *South China Morning Post*, March 29, 2023, https://www.scmp.com/news/china/science/article/3215286/chinese-team-behind-extreme-animal-gene-experiment-says-it-may-lead-super-soldiers-who-survive.

13. Peter Clarke, "Transhumanism and the Death of Human Exceptionalism," *Areo*, March 7, 2019, https://areomagazine.com/2019/03/06/transhumanism-and-the-death-of-human-exceptionalism.

14. Andrés Lomeña, "Transhumanism: Nick Bostrom and David Pearce Talk to Andrés Lomeña," *Literal Magazine* 31 (Winter 2012–2013): 5–8, https://literalmagazine.com/assets/l31-web.pdf.

15. Jonathan Cook, "Should AI Cure Humanity of Its Emotions?" *Medium*, August 7, 2018, https://jonathanccook.medium.com/should-ai-cure-humanity-of-its-emotions-2a3a041428e1.

16. Cook, "Should AI Cure Humanity."

17. Emily A. Partridge et al., "An Extra-Uterine System to Physiologically Support the Extreme Premature Lamb," *Nature Communications* 8 (April 25, 2017): 15112, https://doi.org/10.1038/ncomms15112.

18. Zoltan Istvan, "Transhumanist Science Will Free Women from Their Biological Clocks," *Quartz*, January 14, 2019, https://qz.com/1515884/transhumanist-science-will-free-women-from-their-biological-clocks.

19. Istvan, "Transhumanist Science Will Free Women."

20. Max Roser, "Fertility Rate," Our World in Data, last modified March 2024, https://ourworldindata.org/fertility-rate.

21. Roser, "Fertility Rate."

22. "The Lancet: Dramatic Declines in Global Fertility Rates Set to Transform Global Population Patterns by 2100," Institute for Health Metrics and Evaluation, March 20, 2024, https://www.healthdata.org/news-events/newsroom/news-releases/lancet-dramatic-declines-global-fertility-rates-set-transform.

23. *Merriam-Webster Dictionary*, s.v. "eugenics," accessed July 12, 2024, https://www.merriam-webster.com/dictionary/eugenics.

24. Arthur L. Caplan, Glenn McGee, and David Magnus, "What Is Immoral about Eugenics?" *BMJ* 319, no. 7220 (November 13, 1999): 1284, https://doi.org/10.1136/bmj.319.7220.1284.

25. Paul Lombardo, "Eugenics Sterilization Laws," Eugenics Archive, Dolan DNA Learning Center, Cold Spring Harbor Laboratory, accessed July 12, 2024, http://www.eugenicsarchive.org/html/eugenics/essay_8_fs.html.

26. "Fact Sheet: Eugenics and Scientific Racism," National Human Genome Research Institute, accessed July 12, 2024, https://www.genome.gov/about-genomics/fact-sheets/Eugenics-and-Scientific-Racism.

27. "Unidentified Anomalous Phenomena: Implications on National Security, Public Safety, and Government Transparency," U.S. Government Publishing Office, https://www.congress.gov/118/meeting/house/116282/documents/HHRG-118-GO06-Transcript-20230726.pdf. Transcript of David Grusch's testimony under oath before the House of Representatives (118th Congress) Subcommittee on National Security, the Border, and Foreign Affairs on July 26, 2023.

28. Liguo Zhang et al., "Reverse-Transcribed SARS-CoV-2 RNA Can Integrate into the Genome of Cultured Human Cells and Can Be Expressed in Patient-Derived Tissues," *Proceedings of the National Academy of Sciences of the United States of America* 118, no. 21 (May 25, 2021): e2105968118, https://doi.org/10.1073/pnas.2105968118. Also see: Stephanie Seneff and Gregory Nigh, "Worse Than the Disease? Reviewing Some Possible Unintended Consequences of the mRNA Vaccines against COVID-19," *International Journal of Vaccine Theory, Practice, and Research* 2, no. 1 (March 31, 2021; updated June 16, 2021): 38–79, https://doi.org/10.56098/ijvtpr.v2i1.23.

## CHAPTER SEVEN: Deprogramming

1. Klaus Schwab, "Now Is the Time for a 'Great Reset,'" World Economic Forum, June 3, 2020, https://www.weforum.org/agenda/2020/06/now-is-the-time-for-a-great-reset.

2. Klaus Schwab, "What Is the Fourth Industrial Revolution? by Prof Klaus Schwab," Lee Kuan Yew School of Public Policy, July 13, 2016, YouTube video, https://www.youtube.com/watch?v=7xUk1F7dyvI.

3. Abdallah Alami et al., "Risk of Myocarditis and Pericarditis in mRNA COVID-19-Vaccinated and Unvaccinated Populations: A Systematic Review and Meta-Analysis," *BMJ Open* 13, no. 6 (June 20, 2023): e065687, https://doi.org/10.1136/bmjopen-2022-065687.

4. Jun Shimizu et al., "Reevaluation of Antibody-Dependent Enhancement of Infection in Anti-SARS-CoV-2 Therapeutic Antibodies and mRNA-Vaccine Antisera Using FcR- and ACE2-Positive Cells," *Scientific Reports* 12, no. 1 (September 16, 2022): 15612, https://doi.org/10.1038/s41598-022-19993-w.

5. Seyed Mohammad Hassan Atyabi et al., "Relationship between Blood Clots and COVID-19 Vaccines: A Literature Review," *Open Life Sciences* 17, no. 1 (April 26, 2022): 401–15, https://doi.org/10.1515/biol-2022-0035.

6. Markus Aldén et al., "Intracellular Reverse Transcription of Pfizer BioNTech COVID-19 mRNA Vaccine BNT162b2 in Vitro in Human Liver Cell Line," *Current Issues in Molecular Biology* 44, no. 3 (February 25, 2022): 1115–26, https://doi.org/10.3390/cimb44030073.

7. Pritam Bordoloi, "How AI Is Exposing Our Dark Desires for the World's End," *Analytics India Magazine*, April 14, 2023, updated July 15, 2024, https://analyticsindiamag.com/how-ai-exposing-our-dark-desires-the-worlds-end.

8. Khari Johnson, "How Wrongful Arrests Based on AI Derailed 3 Men's Lives," *Wired*, March 7, 2022, https://www.wired.com/story/wrongful-arrests-ai-derailed-3-mens-lives.

9. Hope Reese, "What Happens When Police Use AI to Predict and Prevent Crime?" *JSTOR Daily*, February 23, 2022, https://daily.jstor.org/what-happens-when-police-use-ai-to-predict-and-prevent-crime.

10. Johnson, "How Wrongful Arrests Based on AI Derailed."

11. Victor Rotaru et al., "Event-Level Prediction of Urban Crime Reveals a Signature of Enforcement Bias in US Cities," *Nature Human Behaviour* 6 (June 30, 2022): 1056–68, https://doi.org/10.1038/s41562-022-01372-0.

12. Nadine Kahil, "AI Can Now Predict Crime before It Happens," *Wired*, July 15, 2022, https://wired.me/technology/ai-can-now-predict-crime-before-it-happens.

13. Kahil, "AI Can Now Predict Crime."

14. Patricia Kosseim, "Privacy and Humanity on the Brink," Information and Privacy Commissioner of Ontario (blog), July 21, 2022, https://www.ipc.on.ca/privacy-and-humanity-on-the-brink.

15. "The Great Reset: 'You'll own nothing and you'll be happy.' (World Economic Forum)," Moisterrific, February 13, 2022, YouTube video, https://www.youtube.com/watch?v=SqzepGBatWo. Video released in June 2020 at the annual World Economic Forum in Davos, Switzerland, encapsulating the vision for the Great Reset of the World.

16. "Half of High School Students Already Use AI Tools," ACT.org, December 11, 2023, https://leadershipblog.act.org/2023/12/students-ai-research.html. ACT is a nonprofit organization designed to prepare students for success in the classroom.

17. Daniel H. Abbott, ed., *The Handbook of 5GW: A Fifth Generation of War?* (Ann Arbor: Nimble Books, 2010), 20.

18. "What Is the Meaning of 'Game'?" Bab.la, accessed July 12, 2024, https://en.bab.la/dictionary/english/game.

19. "Holocaust Victims," Wikipedia, accessed July 12, 2024, https://en.wikipedia.org/wiki/Holocaust_victims.

20. Ernest Harsch, "OAU Sets Inquiry into Rwanda Genocide: A Determination to Search for Africa's Own Truth," *Africa Recovery* 12, no. 1 (August 1998): 4.

21. Immanuel Kant, *Perpetual Peace: A Philosophical Essay*, trans. Mary Campbell Smith (London: George Allen and Unwin, 1917), 161.

22. Saul D. Alinsky, *Rules for Radicals: A Practical Primer for Realistic Radicals* (New York: Vintage, 1989), 130.

23. "Film: The Fog of War: Transcript," https://www.errolmorris.com/film/fow_transcript.html (accessed July 14, 2024).

24. Robert J. Hanyok, "Skunks, Bogies, Silent Hounds, and the Flying Fish: The Gulf of Tonkin Mystery, 2–4 August 1964," Naval History and Heritage Command, accessed July 12, 2024, https://www.history.navy.mil/research/library/online-reading-room/title-list-alphabetically/s/skunks-bogies-silent-hounds-flying-fish.html.

25. Steven Hassan, *Combatting Cult Mind Control* (Rochester, VT: Park Street, 1988).

26. Sivakumaran Sivaramanan, "Global Warming and Climate Change, Causes, Impacts and Mitigation," ResearchGate, September 15, 2015, https://doi.org/10.13140/RG.2.1.4889.7128. See Graph of Temperature vs. $CO_2$.

27. Craig Idso, Keith Idso, and Sherwood B. Idso, "Ice Core Studies Prove $CO_2$ Is Not the Powerful Climate Driver Alarmists Make It Out to Be," *$CO_2$ Science* 6, no. 26 (June 25, 2003), http://www.co2science.org/articles/V6/N26/EDIT.php.

28. Buckminster Fuller, *Operating Manual for Spaceship Earth* (New York: Simon & Schuster, 1969).


## CHAPTER EIGHT: Pure Human


1. James M. Robinson, ed., *The Nag Hammadi Library in English*, trans. the Coptic Gnostic Library Project of the Institute for Antiquity and Christianity (San Francisco: HarperSanFrancisco, 1990), saying 70.

# Index

## A



## B

# C

# D

# E



# F



# G

# H



# I

# J



# K



# L



# M

# Q



# R



# S

# T

# U



# V



# W



# Y



# Z

# About the Author

**GREGG BRADEN** is a five-time *New York Times* best-selling author, scientist, and pioneer in the emerging paradigm bridging science, social policy, and human potential.

From 1979 to 1991, Gregg worked as a problem-solver during times of crisis for Fortune 500 companies, including Martin Marietta (now Lockheed Martin), where he worked as a Senior Computer Systems Designer, and Cisco Systems, where he became the first technical operations manager in 1991. He continues problem-solving today, and his research resulted in the 2003 discovery of intelligent information coded into the human genome and the 2010 application of fractal time to predict future occurrences of past events.

Gregg's work has led to 17 film credits and 12 award-winning books now published in over 40 languages, and he was a 2020 nominee for the prestigious Templeton Prize established by Sir John Templeton to honor "outstanding individuals who have devoted their talents to expanding our vision of human purpose and ultimate reality."

He has presented his discoveries in 34 countries on six continents and has been invited to speak to the United Nations, Fortune 500 companies, and the U.S. military.

**www.greggbraden.com**

**Hay House Titles of Related Interest**

*YOU CAN HEAL YOUR LIFE, the movie,*
starring Louise Hay & Friends
(available as an online streaming video)
www.hayhouse.com/louise-movie

*THE SHIFT, the movie,*
starring Dr. Wayne W. Dyer
(available as an online streaming video)
www.hayhouse.com/the-shift-movie

• • •

*BECOMING SUPERNATURAL: How Common People Are Doing the Uncommon,* by Dr. Joe Dispenza

*BIOLOGY OF BELIEF 10th Anniversary Edition: Unleashing the Power of Consciousness, Matter & Miracles,* by Bruce Lipton, Ph.D.

*WHOLE BRAIN LIVING: The Anatomy of Choice and the Four Characters That Drive Our Life,* by Dr. Jill Bolte Taylor

All of the above are available at your local bookstore,
or may be ordered by contacting Hay House (see next page).

• • •

We hope you enjoyed this Hay House book. If you'd like to receive our online catalog featuring additional information on Hay House books and products, or if you'd like to find out more about the Hay Foundation, please contact:

Hay House LLC, P.O. Box 5100, Carlsbad, CA 92018-5100
(760) 431-7695 or (800) 654-5126
www.hayhouse.com® • www.hayfoundation.org

---

***Published in Australia by:***
Hay House Australia Publishing Pty Ltd
18/36 Ralph St., Alexandria NSW 2015
*Phone:* +61 (02) 9669 4299
www.hayhouse.com.au

***Published in the United Kingdom by:***
Hay House UK Ltd
1st Floor, Crawford Corner,
91–93 Baker Street, London W1U 6QQ
*Phone:* +44 (0)20 3927 7290
www.hayhouse.co.uk

***Published in India by:***
Hay House Publishers (India) Pvt Ltd
Muskaan Complex, Plot No. 3,
B-2, Vasant Kunj, New Delhi 110 070
*Phone:* +91 11 41761620
www.hayhouse.co.in

---